MW00838075

JR. HIGH STUDENT

THE NERVOUS SYSTEM

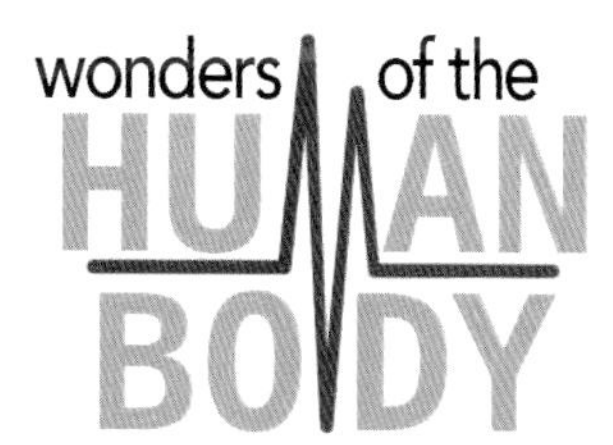

Dr. Tommy Mitchell

First printing: September 2017

Master Books® is a division of the New Leaf Publishing Group, Inc.

ISBN: 978-1-68344-027-7
ISBN: 978-1-61458-616-6 (digital)
Library of Congress Number: 2017908538

Cover by Diana Bogardus
Interior by Jennifer Bauer

Unless otherwise noted, Scripture quotations are from the New King James Version of the Bible.

Please consider requesting that a copy of this volume be purchased by your local library system.

Printed in China

Please visit our website for other great titles:
www.masterbooks.com

For information regarding author interviews, please contact the publicity department at (870) 438-5288.

Dedication

For my dear friends,
Ken and Mally Ham

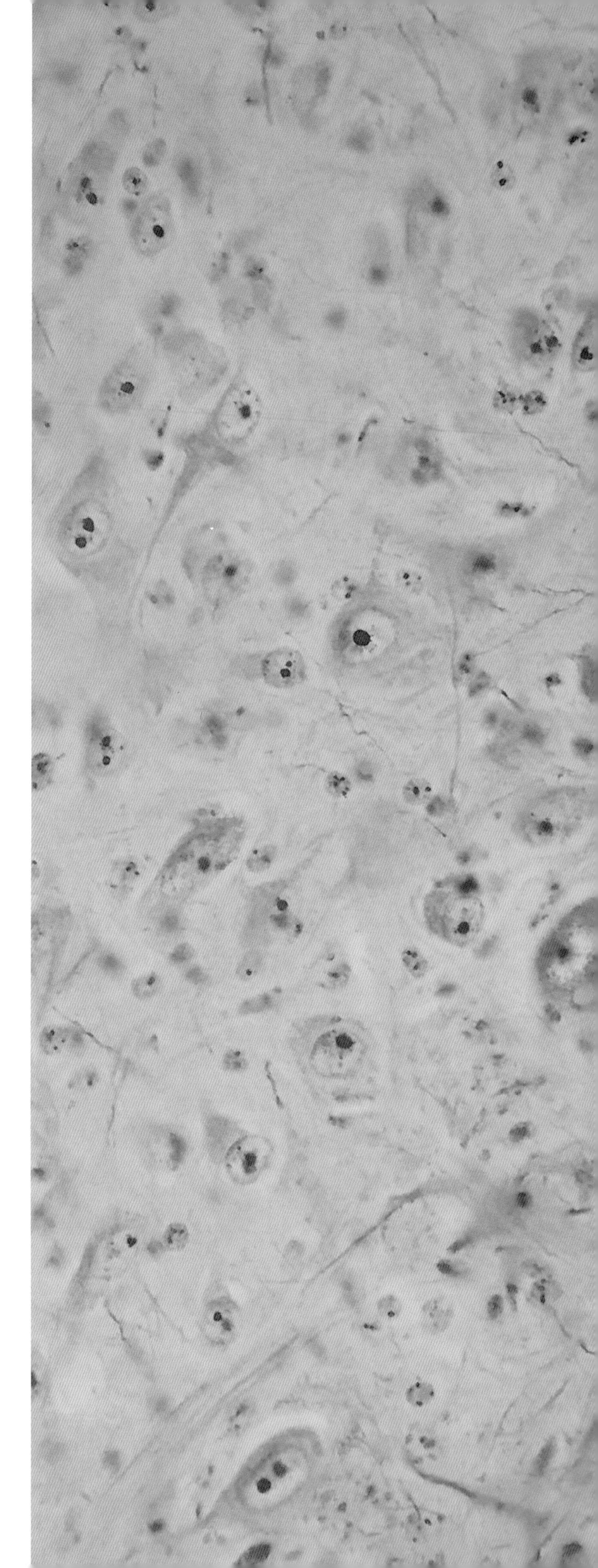

Neurons from the spinal cord under microscope view

TABLE OF CONTENTS

INTRODUCTION

Think for just a moment about the things you do every day. You wake up, walk to the bathroom, comb your hair, and brush your teeth. You sit at your desk and read a book. You take a walk with your dog. You stand in church, and sing a beautiful worship song (hopefully in the right key). You have a conversation with your parents. You go to bed and go to sleep.

How do just the correct muscles know how to contract in just the right way to allow us to walk? How can we control the movements of our hands in a very precise fashion so that we can brush our teeth? How can we decipher those funny marks on a printed page, understand that they are letters and punctuation marks, and make sense of them? How can we hear others singing and make our voices match theirs? How can we understand others' speech? What makes us fall asleep and then wake up again?

Somehow we just "know" how to do these things. Or at least we remember "learning" how to do them. How is this possible? These remarkably complex tasks seem simple because of the remarkably complex human nervous system.

Functions of the Nervous System

The nervous system processes an amazing amount of information. Sometimes this processing is relatively simple, but often it is incredibly complicated. However, as we explore your master control system in more detail, you will notice that all its processes follow the same basic pattern.

This basic pattern is simply this: information comes into the nervous system, this information is recognized and processed, and then a signal is sent out instructing an organ (or organs) to respond in some manner. If you think of the nervous system functioning in this fashion, things won't seem complicated at all.

Let's look at the three parts of this pattern in more detail.

The first step is sensory function. A vast number of sensory receptors throughout the body provide input to the nervous system. There are receptors designed to detect internal changes, such as blood pressure or acid levels in the blood. Other receptors detect external stimuli, such as heat or cold on the skin, or the sensation of a splinter's sharp point. All these receptors send signals to the nervous system. These signals are the sensory inputs.

The next step is called integration. The nervous system integrates all this incoming information. It must recognize, analyze, and process all the

TAKING A CLOSER LOOK

Sensory vs Motor

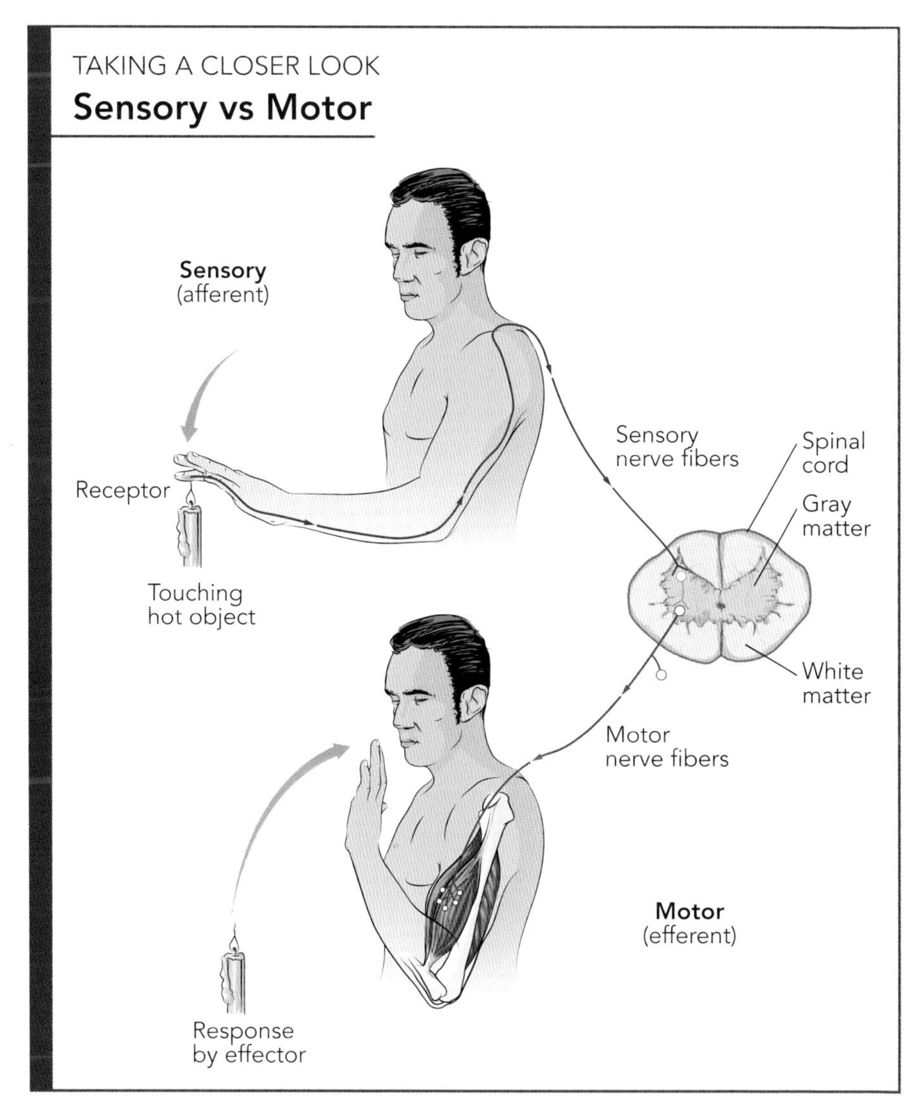

various sensory inputs, often comparing what is sensed in the present to what has been experienced in the past. Then the nervous system comes up with an appropriate response, sometimes filing the information away for future use and often creating an instruction to be sent out to deal with the information.

TAKING A CLOSER LOOK

Central & Peripheral Nervous System

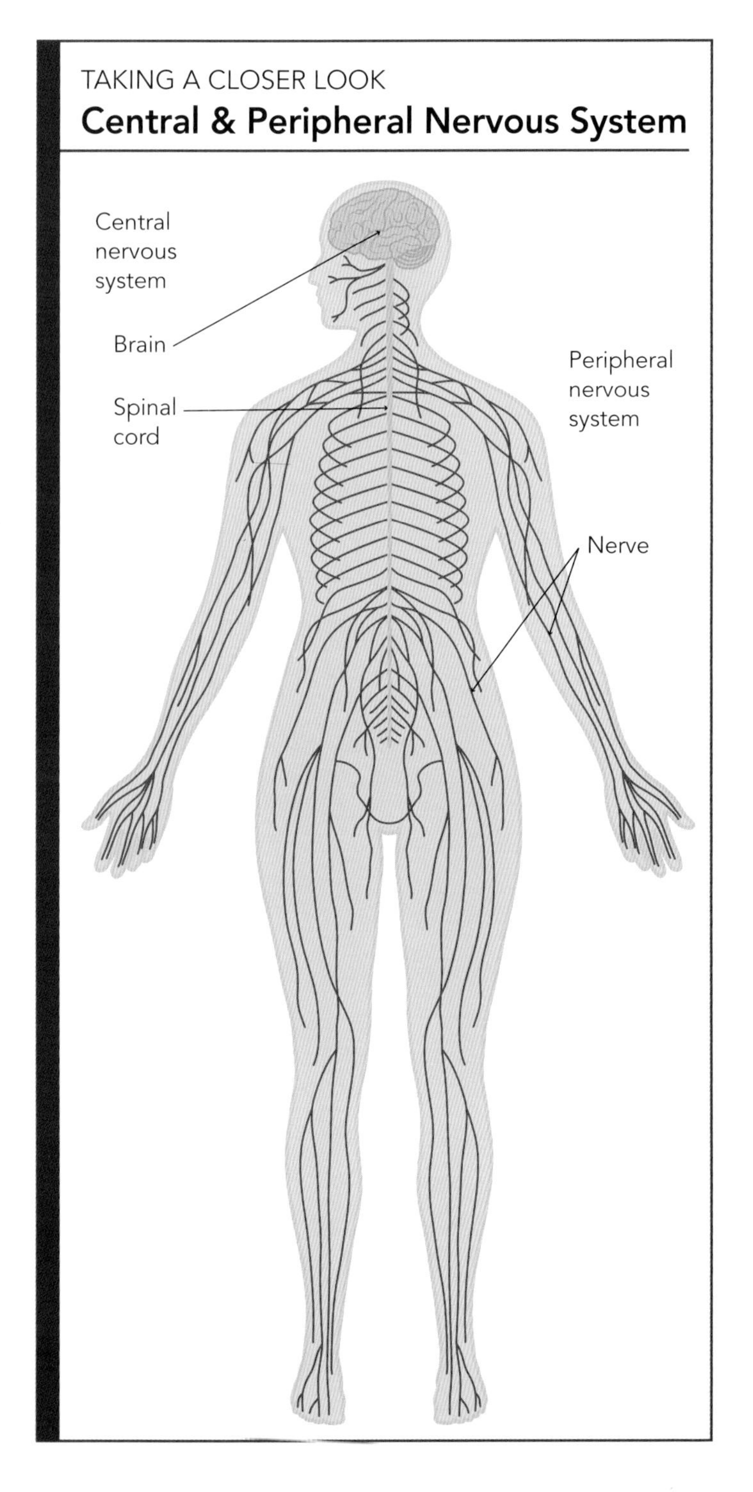

As an example, let's say that you are riding your bike down a steep hill. You feel the wind on your face and sense the speed of the bike increasing. While processing these sensory inputs, you also remember that last month you were going too fast down this hill, wrecked your bike, and sprained your wrist. The processing of sensory inputs is often dependent on your past knowledge and experiences. As your nervous system integrates all this information, you realize that you need to slow down.

The last step is motor output. The word *motor* implies movement or some sort of action. Motor output is simply what the body is told to do as the result of all this information input and processing. In our example, this step causes you to use the muscles in your legs or hands to put some pressure on your coaster brakes or hand brake, and you slow down to a safer speed.

Input, integration, output. Using these three steps, the nervous system controls the complex activities of the human body.

Overview of the Nervous System

We will begin our tour of the nervous system by taking a broad look at its two major divisions, the central nervous system (CNS) and the peripheral nervous system (PNS). Even though these parts work together as a highly efficient, integrated unit, breaking it down into these two parts can be very helpful as we try to understand how the nervous system works.

The central nervous system is composed of the brain and the spinal cord. The brain is the most recognizable part of the CNS. It is the master control center of the nervous system, containing hundreds of millions of neural connections. Our perception of the world around us, our movements, our intellect, our memories—all are controlled and regulated by the brain. The spinal cord extends from the base of the brain down to the lower levels of the spinal column. It provides a pathway for nerves to and from the brain.

The peripheral nervous system is the portion of the nervous system outside of the central nervous system. It consists of the cranial nerves that extend from the brain, and the spinal nerves that extend from the spinal cord. The peripheral nervous system in effect allows all the other organ systems and body parts to connect and interact with the central nervous system.

The PNS has two basic functions: carrying sensory information to the CNS and transmitting instructions out to the various part of the body. Based on these functions, we can divide the PNS into two divisions, the sensory division and the motor division.

The sensory division carries information from the skin and muscles as well as from the major organs in the body to the central nervous system, where all the sensory input is processed ("integrated"). The sensory division is sometimes called the afferent (meaning "bringing toward") division because it carries nerve impulses "to" or "toward" the CNS.

The motor division, on the other hand, carries instructions from the CNS out to the body. (This is the motor output function of the nervous system discussed earlier.) The motor division is sometimes called the efferent (meaning "carrying away") division because it carries instructions "away from" the CNS.

Some instructions carried by the motor division are taken to muscles that we can consciously control. For example, we can consciously control the muscles we use to hold a glass or throw a ball. This aspect of the motor division is called the somatic nervous system.

TAKING A CLOSER LOOK

Sensory vs Motor Nerves

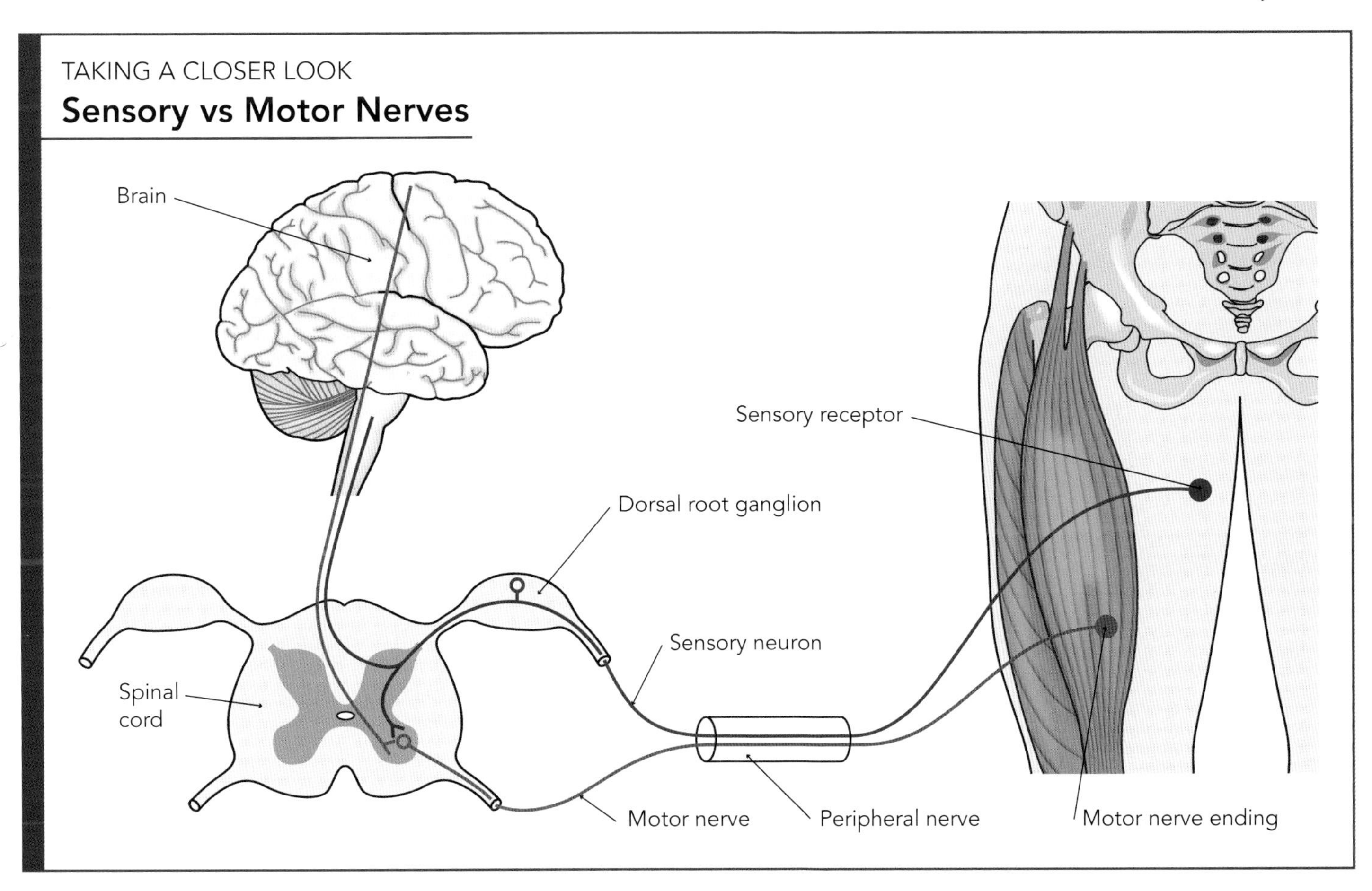

TAKING A CLOSER LOOK

Automatic Nervous System

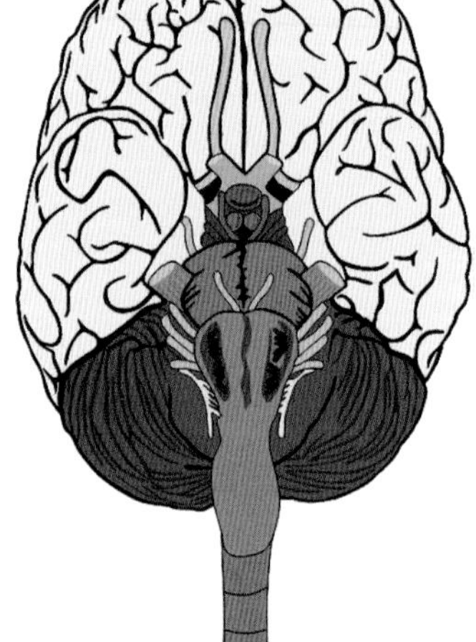

PARASYMPATHETIC

Eye
Constricts pupil

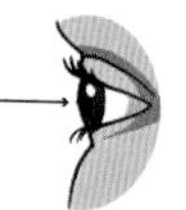

Salivary & Parotid Glands
Stimulates saliva production

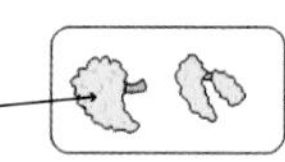

Blood Vessels
Constricts blood vessels in skeletal muscles

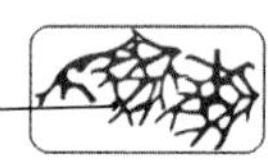

Sweat Gland
Inhibits sweat secretion

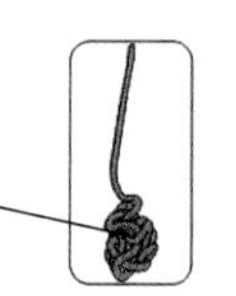

Lungs
Constricts bronchi

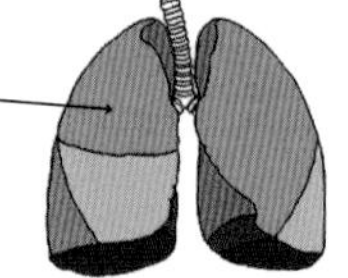

Heart
Slows heart beat

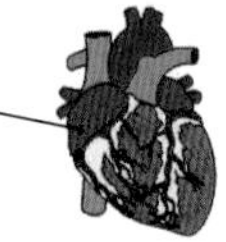

Liver
Inhibits glucose release

Gallbladder
Stimulates bile

Pancreas
Stimulates pancreas

Stomach
Stimulates stomach motility & secretions

Intestines
Stimulates intestinal motility

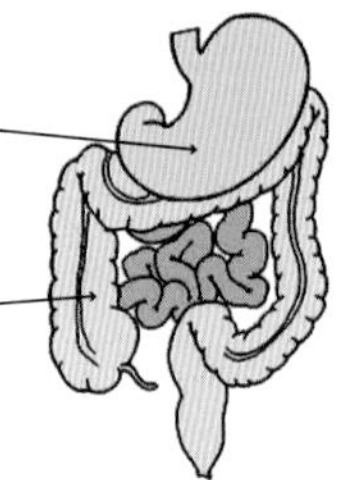

Kidneys
Decreases renin secretion (lowers blood pressure)

Bladder
Stimulates urination

SYMPATHETIC

Eye
Dilates pupil

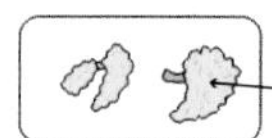

Salivary & Parotid Glands
Inhibits saliva production

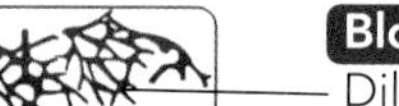

Blood Vessels
Dilates blood vessels in skeletal muscles

Sweat Gland
Stimulates sweat secretion

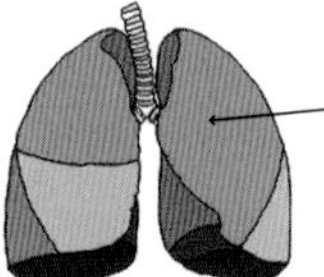

Lungs
Dilates bronchi

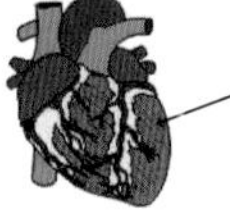

Heart
Accelerates heart beat

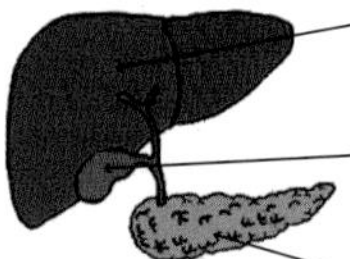

Liver
Stimulates glucose release

Gallbladder
Inhibits bile

Pancreas
Inhibits pancreas

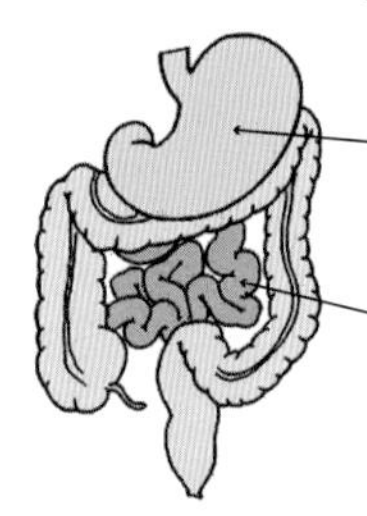

Stomach
Inhibits stomach motility & secretions

Intestines
Inhibits intestinal motility

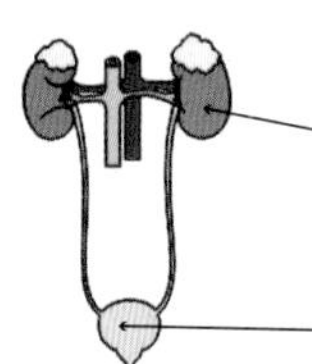

Kidneys
Increases renin secretion (raises blood pressure)

Bladder
Inhibits urination

Somatic means "body," so the somatic nervous system allows us to control our body's movements.

Another equally important part of the nervous system's motor division controls involuntary activities. Involuntary activities—like making sure we breathe and adjusting our heart rate—are vital to survival but not under voluntary control. Such activities continue 24 hours a day whether we think about them or not, and that is a very good thing. Imagine having to think about every single breath you take! What would happen when you slept? It is a good thing the nervous system takes care of this for us. The part of the motor division that controls these involuntary functions is called the autonomic nervous system. (Autonomic sounds a lot like "automatic," so you should be able to remember this easily!)

Let's make sure you have all these divisions and subdivisions straight so far. The nervous system has two parts: the central nervous system and the peripheral nervous system. The central nervous system consists of the brain and spinal cord. The peripheral nervous system brings information to the central nervous system with its sensory nerves, and it transmits instructions from the central nervous system with its motor nerves. Somatic motor nerves instruct skeletal muscles to move voluntarily. Autonomic motor nerves carry instructions for involuntary functions, like breathing and adjustments of the heart rate.

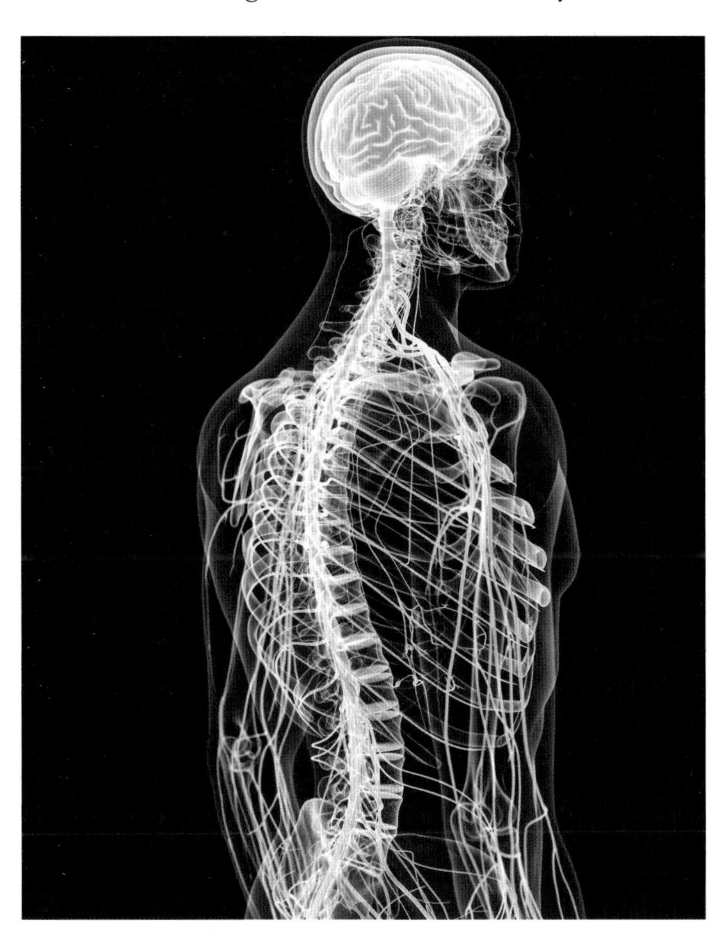

The autonomic nervous system also consists of two parts: the sympathetic nervous system and the parasympathetic nervous system. Both control our involuntary functions, but they have opposite effects on the body. The sympathetic division is more active when we are stressed or exercising. Think of the sympathetic nervous system as the part of you that triggers your "fight or flight" responses to danger. The parasympathetic division does the opposite. The parasympathetic nervous system promotes less-demanding activities like digestion, things that your body needs to do while not busy running or expending lots of energy on other highly active pursuits. Both sympathetic and parasympathetic functions are important for the body to operate properly.

If at this point you are feeling a little overwhelmed with all this, don't worry. Everyone feels that way the first time they encounter all these "divisions." Just keep sight of the big picture and everything will soon fall into place. Remember the three basic functions of the nervous system? They are sensory input, integration, and motor output. No matter how bewildering all these divisions seem to be right now, it all comes down to the basic three functions.

As we examine the nervous system in more detail, you will see just how sensibly it is organized. And you will be amazed at how it works as it assists and controls complex activities throughout your body.

2

STRUCTURE OF NERVOUS TISSUE

The nervous system is composed primarily of nervous tissue. Nervous tissue is one of the four basic tissue types that we examined previously in Volume 1 of *Wonders of the Human Body*.

Nervous tissue consists of two primary types of cells: neurons and neuroglia.

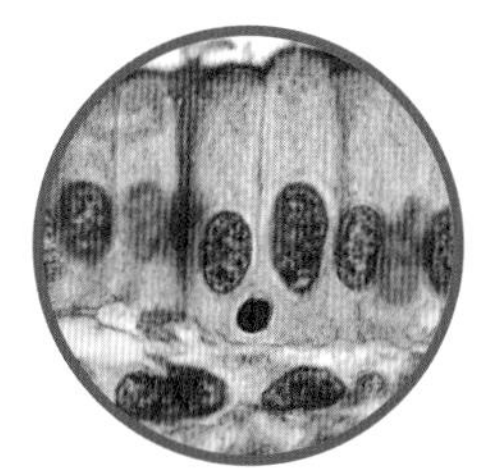

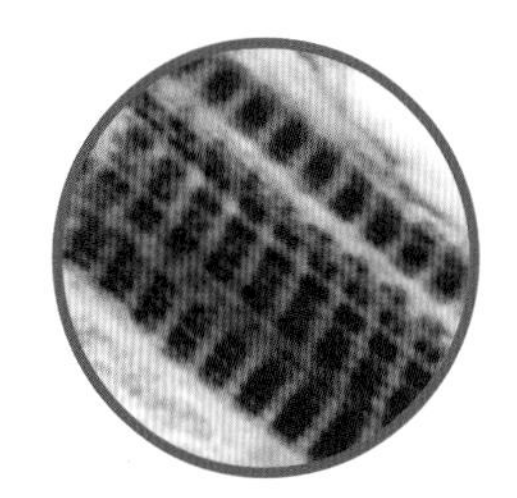

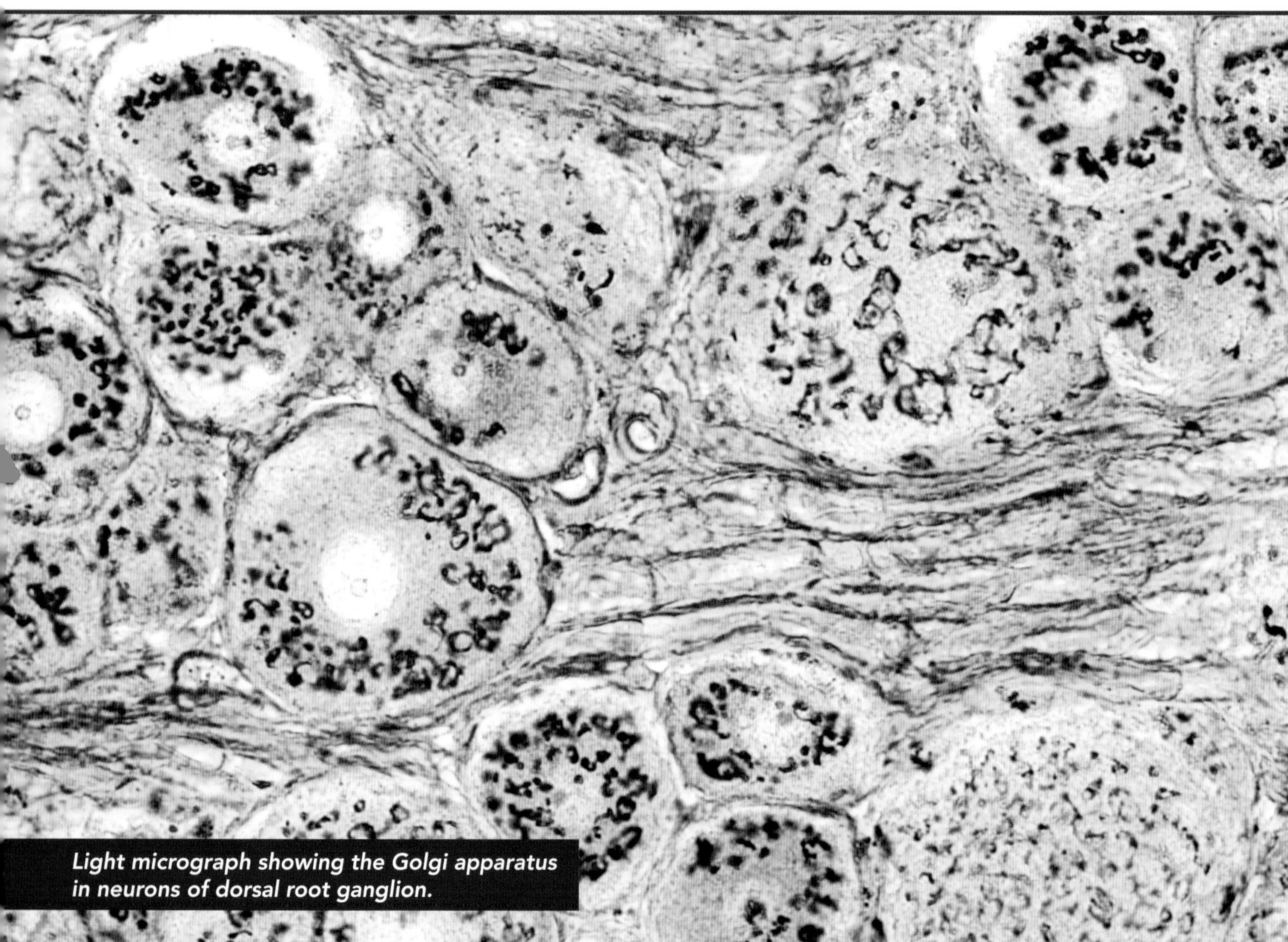

Light micrograph showing the Golgi apparatus in neurons of dorsal root ganglion.

Tissue Types

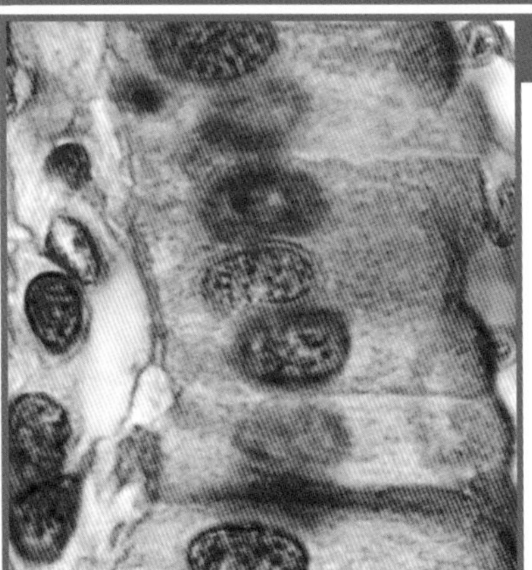

Epithelial Tissue

Epithelial tissue (or epithelium) lines body cavities or covers surfaces. For example, the outer layer of skin is epithelium. The sheet of cells that line the stomach and intestines, as well as the cells that line the heart, blood vessels, and the lungs, is epithelial tissue.

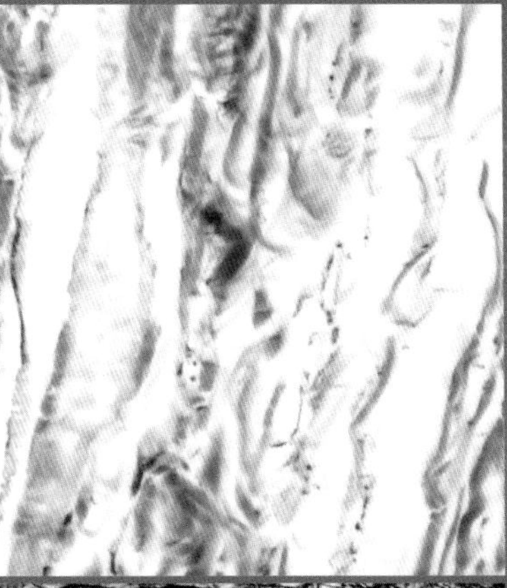

Connective Tissue

Connective tissue helps provide a framework for the body. It also helps connect and support other organs in the body. Further, it helps insulate the body, and it even helps transport substances throughout the body. This tissue can be hard or soft. Some connective tissue stretches. One type is even fluid. Connective tissue is comprised of three parts: cells, fibers, and ground substance.

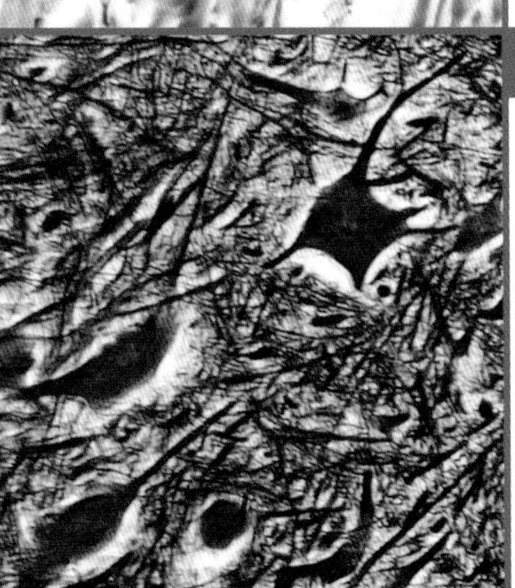

Nervous Tissue

Nervous tissue is the primary component of the nervous system. The nervous system regulates and controls bodily functions.

Nerve cells are incredible. They are able to receive signals or input from other cells, generate a nerve impulse, and transmit a signal to other nerve cells or organs.

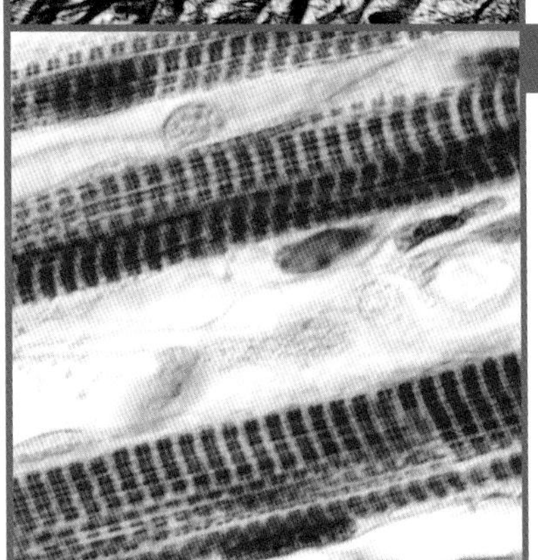

Muscle Tissue

Muscle tissue is responsible for movement. There are three types of muscle tissue: skeletal muscle, smooth muscle, and cardiac muscle.

Neurons are the excitable nerve cells that transmit electrical signals.

What starts such an electrical signal? Some type of change in the environment acts as the stimulus that excites a neuron, triggering an electrical signal called an action potential. The electrical signal transmitted by a neuron is also called an impulse. An impulse travels like a wave along the nerve cell membrane from one end of the neuron to another. We will soon study this in depth.

The other cells in nervous tissue are called neuroglia. There are several types of neuroglia cells. They help protect and support the neurons.

Let's examine the neuron in greater detail.

Neurons

The neuron is often called a nerve cell because it is the cell type that does the primary work of the nervous system. You have neurons in your brain, in your spinal cord, in your peripheral nervous system, and even in specialized sensory organs like your eye, nose, and ear.

A neuron doesn't look like a typical cell. If you have seen sketches of "typical" cells before, you will notice that, while the neuron still has a cell membrane, cytoplasm, and a nucleus, it has an unusual shape. The neuron is a very specialized type of cell that is designed to transmit electrical impulses (nerve impulses) rapidly to various parts of the body.

The neuron is composed of three parts: the cell body, dendrites, and the axon.

The cell body contains the typical organelles we discussed at length in Volume 1 of *Wonders of the Human Body*. The cell body contains a nucleus surrounded by cytoplasm. The cytoplasm contains plenty of protein-building organelles like rough endoplasmic reticulum dotted with ribosomes and free ribosomes. An extensive Golgi apparatus processes the proteins made by these ribosomes. Neurons require a lot of energy to build the substances they require, so lots of energy-generating mitochondria are also found in the cell body. Energy provided by these mitochondria fuels the building of the substances neurons need to do their job. Some of the most important substances synthesized in the neuron's cell body are neurotransmitters. As we will soon see, neurotransmitters are the chemicals that transmit an electrical impulse from one neuron to the next.

Extending from the cell body are numerous projections, or processes. Neuron cell bodies have two kinds of processes protruding from them, dendrites and axons. Dendrites are designed to receive signals. Axons are designed to carry signals away.

Some dendrites resemble the branches of a tree. Others have more thread-like branches, and some have branches covered with tiny spines. The reason for this branching design is simple. Remember, dendrites are the parts of neurons that receive inputs (signals). The branching pattern covers an extensive area, allowing the neuron to receive an enormous number of inputs. When an input is received by a dendrite, an electrical signal is generated and transmitted toward the cell body.

The axon is the portion of the neuron that carries a nerve impulse away from the cell body. The axon begins at a cone-shaped axon hillock on the cell body. The hillock narrows to form the more thread-like axon. The axon can be very short or up to several feet long. The axon of a motor nerve to the muscle that enables you to curl your big toe has to travel a long way, all the way from your spinal cord to your foot.

TAKING A CLOSER LOOK

Neuron

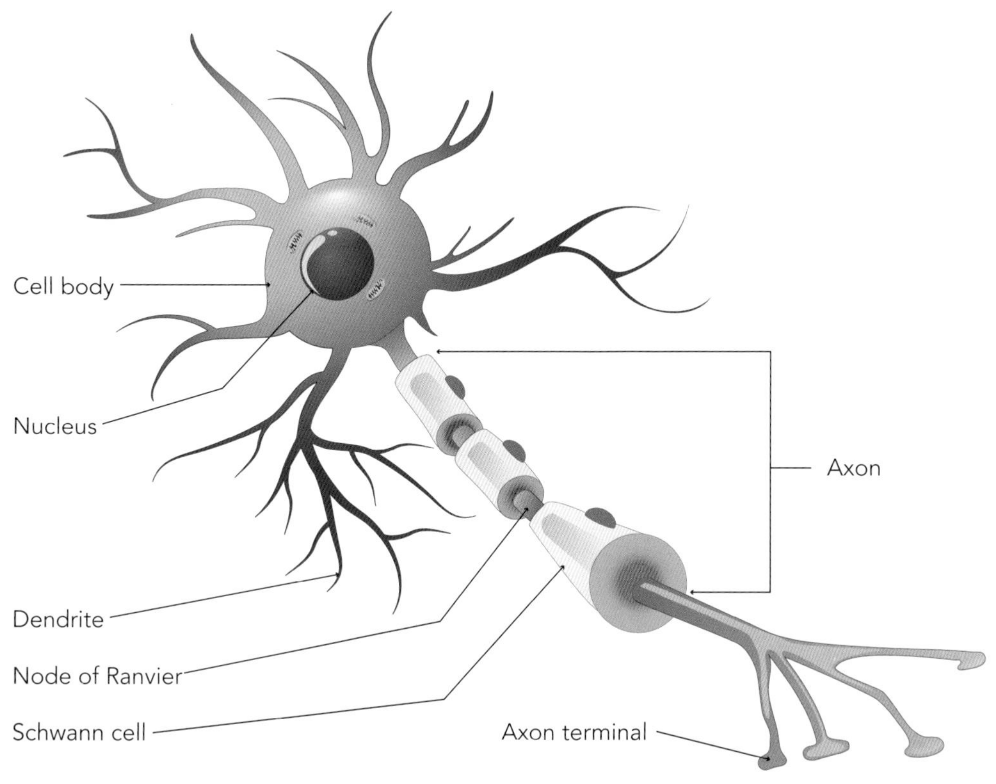

A neuron can have multiple dendrites but only one axon. Axons end in small branches called axon terminals. At the axon terminal, neurotransmitters are released to carry the neuron's signal on to the next cell in line. You will learn more about this shortly.

Neurons — The Lowdown

There are hundreds of millions of neurons in the human body. And that's a really good thing. Why? Unlike most cell types in your body, neurons cannot be routinely replaced. Once neurons mature, with only rare exceptions, they are no longer able to divide. The neurons you have, once your nervous system matures, are all the neurons you will ever have.

So...when neurons are damaged by drugs, disease, or injury, the loss of function is often permanent. Neurons are designed to last a lifetime, but we need to take care of them. For instance, we must be vigilant about what we put into our bodies, as many illicit drugs destroy these precious messengers. A lifetime of poor eating habits and lack of exercise can increase the risk of a stroke in later life, which can destroy many neurons in the brain. Riding your bicycle without a helmet puts the irreplaceable neurons in your brain at risk right now. Following the rules for safety in contact sports may prevent a tragic accident that could leave you paralyzed. Operating power tools unsafely may lead to permanent loss of peripheral nerve function in an injured body part, even if you do not lose the body part itself. Habitually exposing your ears to loud music or explosive noise without ear protection may destroy the specialized neural structures in your ears and impair your hearing. Looking directly at the sun can permanently damage your retina, the very specialized extension of your brain that enables you to see.

God only gave you one body, and there are no do-overs when it comes to neuron damage. While many diseases and conditions that damage neurons in this sin-cursed world are not preventable, you should take care to avoid those that are.

Further, neurons require lots of oxygen and glucose to function properly. Neuron cells can be quickly damaged by lack of these essentials. Loss of oxygen for as little as four minutes can permanently damage neurons. For this reason, many people take courses in basic CPR and water safety, so that they will be able to help others avoid permanent damage or loss of life.

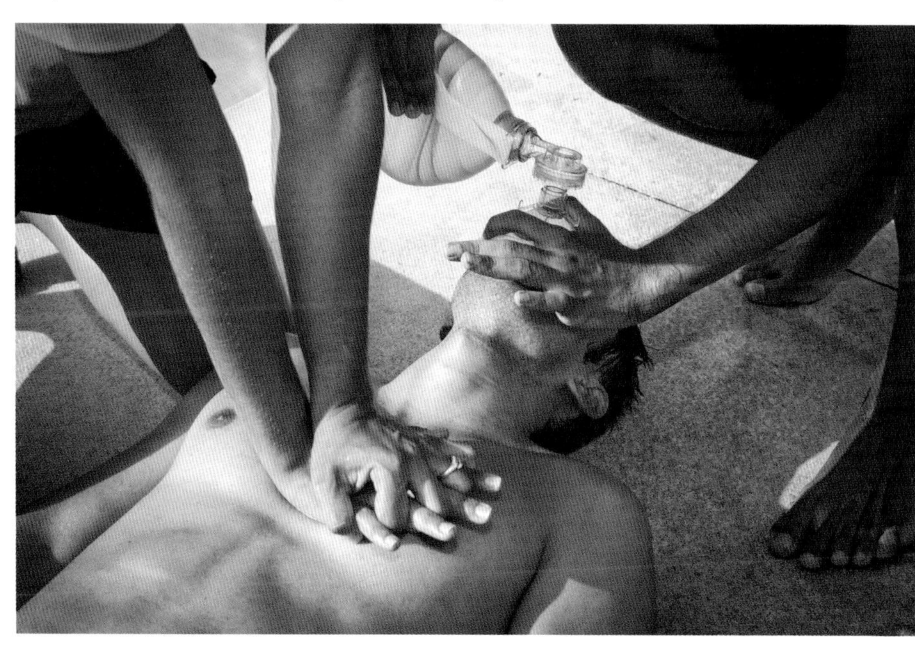

Performing CPR (cardiopulmonary resuscitation) on someone who has stopped breathing.

Types of Neurons

There are several types of neurons. We can classify them according to how they look or according to how they work. Each type of classification can help us understand how the nervous system works.

One method of classifying neurons is based on the number of processes they have. Remember, processes are dendrites and axons, the projections sticking out from the cell body.

Most neurons have one axon and multiple dendrites. These are called multipolar neurons. This is by far the most common type of neuron in the body.

Bipolar neurons have only two processes: one axon and one dendrite. These are only found in special sensory organs, such as the eye, ear, and nose.

Unipolar neurons have a more unusual configuration. They have only one process extending from the cell body. This process looks like a "T." The dendrite and the axon form the arms of this "T."

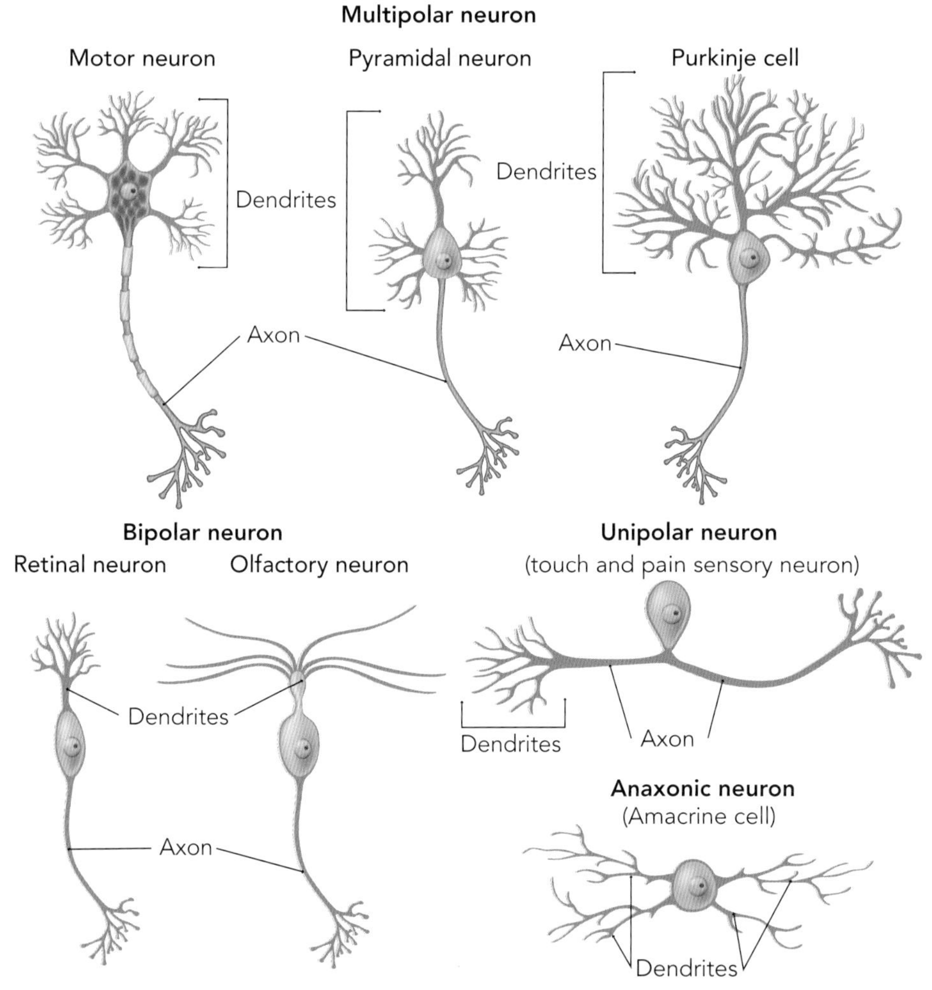

Neurons are also classified according to the direction they carry nerve impulses. Some neurons carry instructions from the central nervous system, and others bring information to the central nervous system.

Neurons that transmit impulses away from the central nervous system are called motor or efferent (remember "carrying away" or "carrying outward") neurons. These impulses contain instructions to muscles or to glands in the body. Most motor neurons are multipolar.

Sensory or afferent (remember "bringing toward") neurons carry impulses triggered by sensory receptors toward the central nervous system. Most sensory neurons are unipolar.

TAKING A CLOSER LOOK

Types of Neurons

Yet one other class of neurons carries impulses from one neuron to another within the central nervous system. These connectors are called interneurons, a word that obviously means "between neurons." Interneurons make up the vast majority of the neurons in the body. Some estimates are as high as 99 percent. Interneurons are located in the brain and spinal cord, forming connections between sensory and motor neurons. Signals from sensory neurons are delivered to the interneurons. The interneurons pass the impulse on to the appropriate motor neurons. If you recall the basic functions of the nervous system, this is the integration step we discussed, a step in which inputs are processed and passed on to generate suitable output.

Neuroglia

Neurons are not usually alone. They are generally surrounded by several types of smaller cells in the nervous system. These other cells are known as neuroglia, or glial cells. Neuroglia are found both in the central nervous system and the peripheral nervous system. Neuroglia have various functions depending on their cell type and location.

We will first examine the neuroglia in the CNS.

Astrocytes are the most numerous of the neuroglial cells in the CNS. Astro means "star," and cytes means "cells." Astrocytes are therefore glial cells with

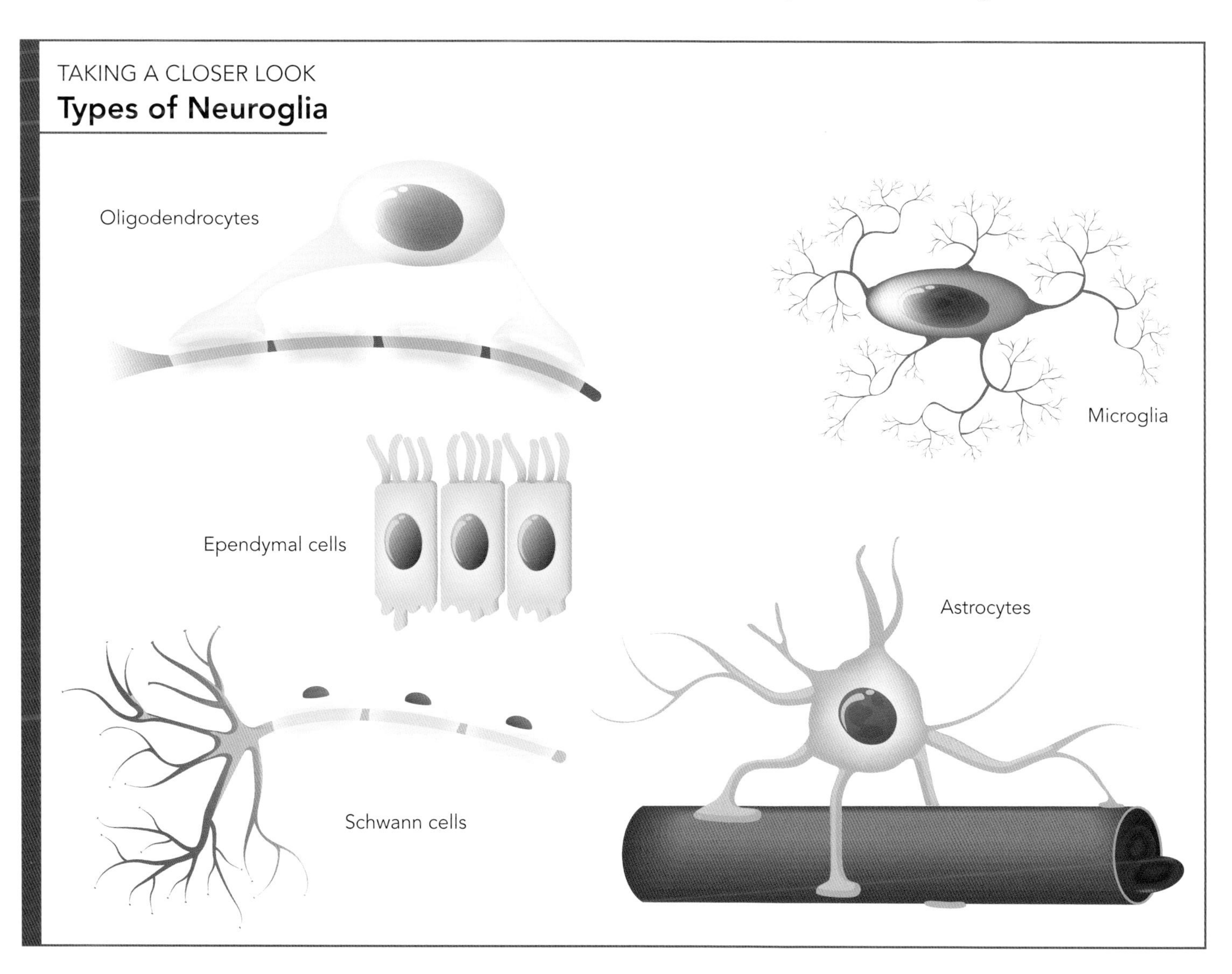

many star-shaped processes. These cells anchor and support the neurons associated with them. They help the neurons pass on impulses efficiently. Astrocytes also protect their neurons. They monitor nearby capillaries, ensuring that harmful substances in the blood do not reach the neuron. Astrocytes help maintain the correct level of ions, such as potassium (K^+), and other nutrients around the neurons. They contain a readily available supply of glucose that they supply to neurons when lots of energy is needed. They even help recycle neurotransmitters released from their neurons.

Microglia are small cells with long slender processes. (Micro means "small," so this is a good name.) Microglial cells "keep watch" over neurons in their vicinity. If they detect damage to a neuron or invading bacteria, they transform into a cell that can remove damaged nerve tissue or engulf and destroy the bacteria.

Ependymal cells line the ventricles of the brain and the spinal canal. The ventricles in the brain, like the canal surrounding the spinal cord, are filled with cerebrospinal fluid. Ependymal cells produce much of the cerebrospinal fluid that fills these cavities. Cerebrospinal fluid doesn't just sit still; it circulates through these fluid-filled spaces in the CNS. Cilia on the ependymal cells help move this fluid around.

Oligodendrocytes resemble astrocytes, but they are smaller. Oligodendrocytes produce and maintain a special covering (called a myelin sheath) around neuronal axons. This myelin sheath is made of lipids and protein. We will be learning much more about myelinated axons shortly.

Okay, now you know there are four types of glial cells in the central nervous system—astrocytes, microglial cells, ependymal cells, and oligodendrocytes. There are two types of neuroglial cells in the peripheral nervous system, satellite cells and Schwann cells.

Satellite cells surround the cell bodies of neurons in the PNS. They provide structural support and also control the extracellular environment around the cell bodies. Thus, the satellite cells function in the PNS much in the way astrocytes do in the CNS.

Schwann cells form the myelin sheaths around axons in the PNS. Therefore, Schwann cells function in the PNS the way oligodendrocytes do in the CNS. Let's explore myelination in more detail next.

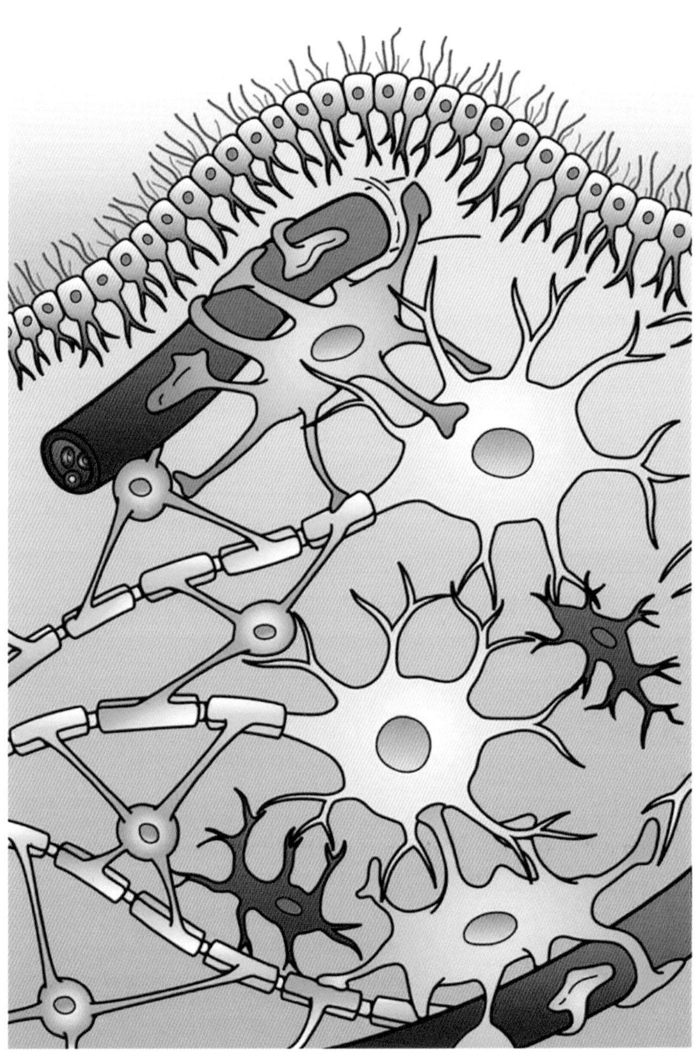

This image shows the four different types of glial cells found in the central nervous system: Ependymal cells (light pink), Astrocytes (green), Microglial cells (red), and Oligodendrocytes (functionally similar to Schwann cells in the PNS) (light blue).

Myelination

Myelination is a process in which long axons are covered by a myelin sheath. The myelin sheath is a spiral wrapping of the modified cell membranes of the Schwann cells or oligodendrocytes responsible for forming the myelin. Axons having this myelin covering are said to be myelinated. Axons not having this covering are called nonmyelinated.

The myelin sheath provides electrical insulation for the axon. It also increases the speed a nerve signal can travel.

In the PNS, myelination is carried out by Schwann cells. These cells initially indent to receive the axon, and then wrap themselves repeatedly around the axon. Ultimately, this wrapping has the appearance of tape wrapped around a wire or gauze wrapped around a finger. At the end of the wrapping process, there may be several dozen layers of wrapping to the sheath.

Each of the Schwann cells wraps only a small length of a single axon. Other Schwann cells wrap the remaining length of the axon, like so many hot dogs in buns laid end to end. However, Schwann cells do not touch each other. There are small gaps between adjacent Schwann cells. These gaps are called nodes of Ranvier. (They were discovered by—you guessed it!—French anatomist Louis-Antoine Ranvier in the 19th century, and his name is pronounced ron'- vee-ay.)

TAKING A CLOSER LOOK

Myelination

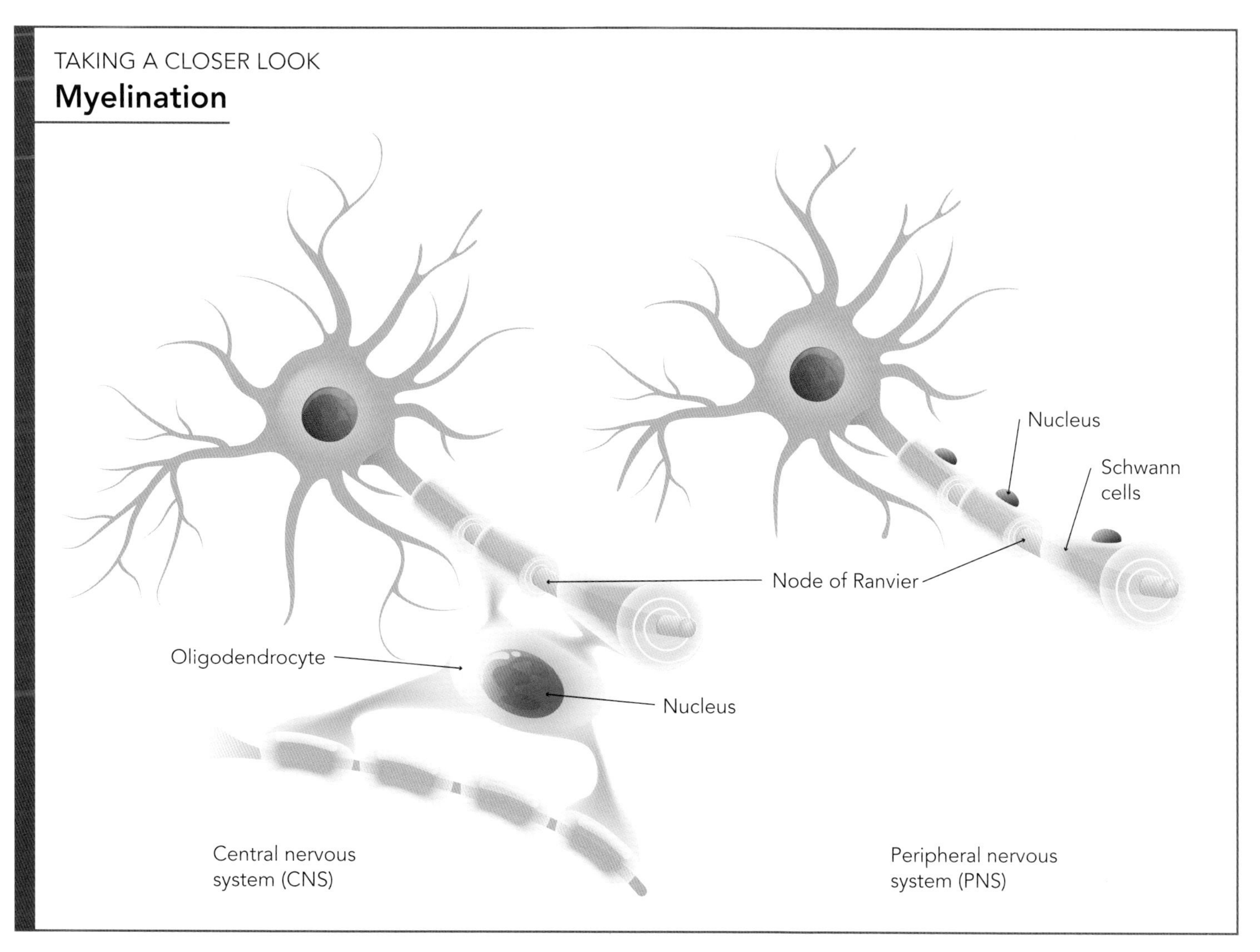

It should be pointed out here that a Schwann cell can enclose a dozen or more axons without wrapping them. These axons are nonmyelinated even though they are in contact with a Schwann cell.

In the CNS, it is the oligodendrocyte that is responsible for myelination. Because an oligodendrocyte has many processes, it can wrap around numerous axons rather that only one, as in the case of the Schwann cell.

The amount of myelin in the body is very low at birth and increases as the body develops and matures. Thus the number of myelinated axons increases from birth throughout childhood until adulthood. Myelination increases the speed of nerve impulse conduction through the axon. Faster conduction

Multiple Sclerosis

Multiple Sclerosis (MS) is an autoimmune disease that results in the destruction of myelin sheaths in the central nervous system. (In autoimmune diseases, the body's immune system turns against its owner's own tissues.) In multiple sclerosis, the body's immune system attacks myelin proteins, creating hardened lesions called scleroses. These lesions commonly occur in the optic nerve, the brain stem, and the spinal cord.

As the myelin loss increases, conduction of nerve impulses becomes progressively slower. Short circuits develop and interfere with the proper functioning of the neurons. That this disease is so debilitating shows the importance of the myelination of nerve fibers to proper functioning of the nervous system.

MS primarily occurs in people under 50 years of age. Symptoms include double vision, weakness, loss of coordination, and paralysis.

One form of MS is characterized by periods of active disease alternating with periods of minimal symptoms. Another form of MS is slowly progressive, without the symptom-free periods.

Although in recent years much progress has been made in our understanding of multiple sclerosis, at present there is still no cure.

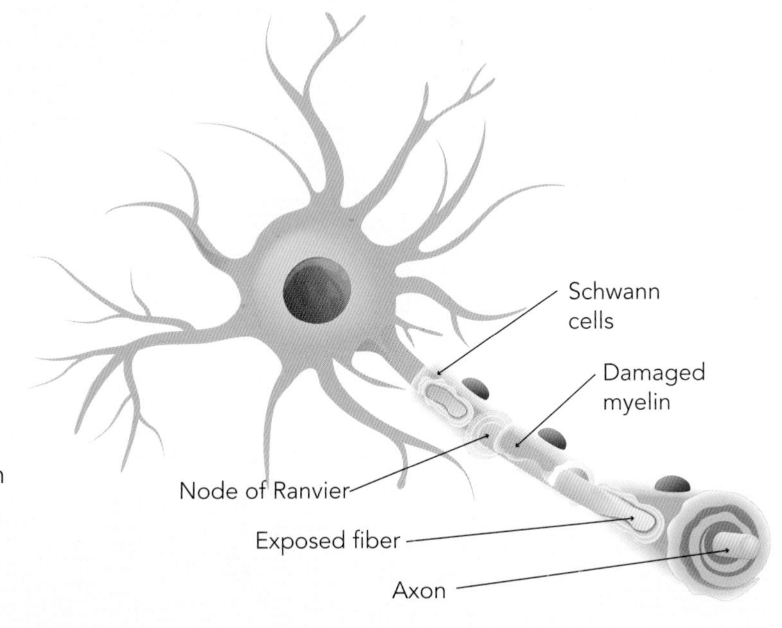

makes those nerves work better, more efficiently, as an individual matures.

Think of a newborn baby. It has very little control of its body in the beginning. It cannot hold its head up or sit up or walk. As more axons become myelinated, it has more and better control of its muscles. Compare this to a teenager. After years of development, the teenager has much better control and coordination of the body. Much of this improvement of due to increased myelination both in the central and peripheral nervous systems.

Nerves

What are nerves? They not the same thing as neurons.

A neuron is a nerve cell. Neurons have dendrites and axons. The neuron is the cell that transmits electrical impulses in the nervous system. Thus, the neuron, not the nerve, is the basic unit of nervous tissue.

So what is a nerve? Well remember that axons, even though they are part of nerve cell, can be very long. Some reach from your back to your foot. A nerve is made of bundles of axons located in the peripheral nervous system. These bundles of axons are not alone in the nerve. The nerve also contains the Schwann cells associated with the axons, as well as blood vessels, connective tissue, and lymphatic vessels. This cross section shows the various components.

Before we go further, let's see how some of these things fit together.

Individual axons and their associated Schwann cells are covered by a very thin layer of connective tissue known as the

TAKING A CLOSER LOOK

Anatomy of a Nerve

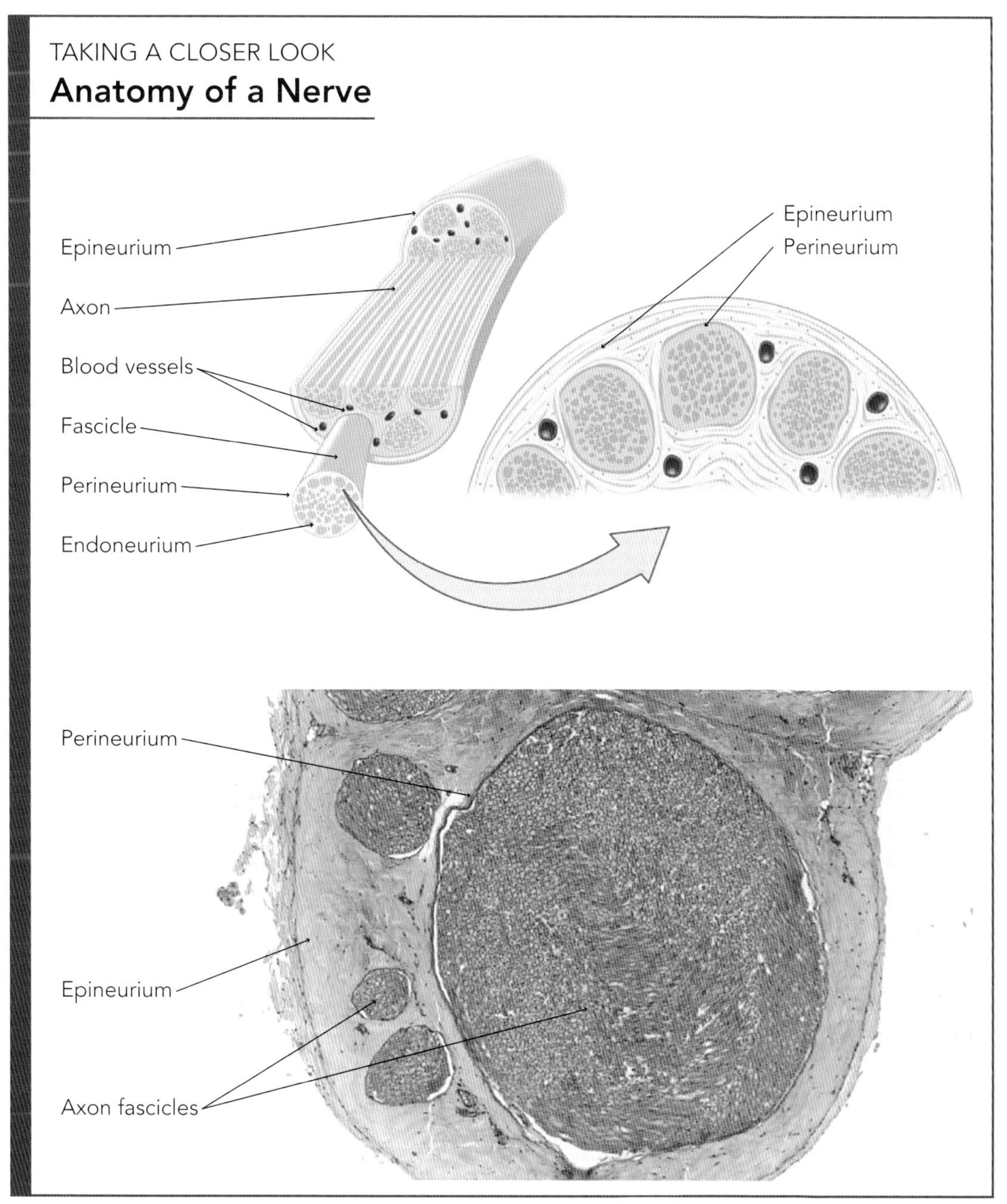

endoneurium (endo- meaning "inner," and neurium meaning "nerve"). Next, many such endoneurium-covered axons running parallel to each other are grouped in bundles called fascicles. Each fascicle is then covered by another connective tissue layer known as the perineurium (peri- meaning "around"). Lastly, numerous fascicles, blood and lymphatic vessels are bound together by yet another connective tissue wrapping called the epineurium (epi- meaning "over"). This epineurium-wrapped bundle of bundles—containing axons, neuroglia, blood vessels, lymphatic vessels, and layers of connective tissue—is known as a "nerve."

Remember that neurons can be classified by the direction they carry electrical impulses. Motor neurons carry impulses away from the central nervous system, and sensory neurons carry impulses toward the central nervous system. Nerves can be classified the same way.

Motor nerves carry signals away from the CNS. Sensory nerves carry impulses toward the CNS. But motor nerves and sensory nerves are very rare. The most common type of nerve by far is called a mixed nerve. Even though an individual neuron can only carry an impulse in one direction (remember, from dendrite to cell body to axon), mixed nerves possess both motor and sensory fibers. Mixed nerves have two-way traffic. They carry impulses both toward and away from the CNS.

Nerve Damage and Repair

With rare exceptions, mature neurons do not divide to reproduce themselves. The mature nervous system is not designed to replace damaged nerve cells. The neurons you have now are pretty much all you are going to get. Because of this, damage to the nervous system is serious.

However, there is a bright spot here. In the peripheral nervous system, there can be regeneration of a nerve after an injury. Recall that a nerve does not contain whole neurons, but instead consists of bundles of axons and their supporting tissues.

TAKING A CLOSER LOOK

Neuron Damage and Repair

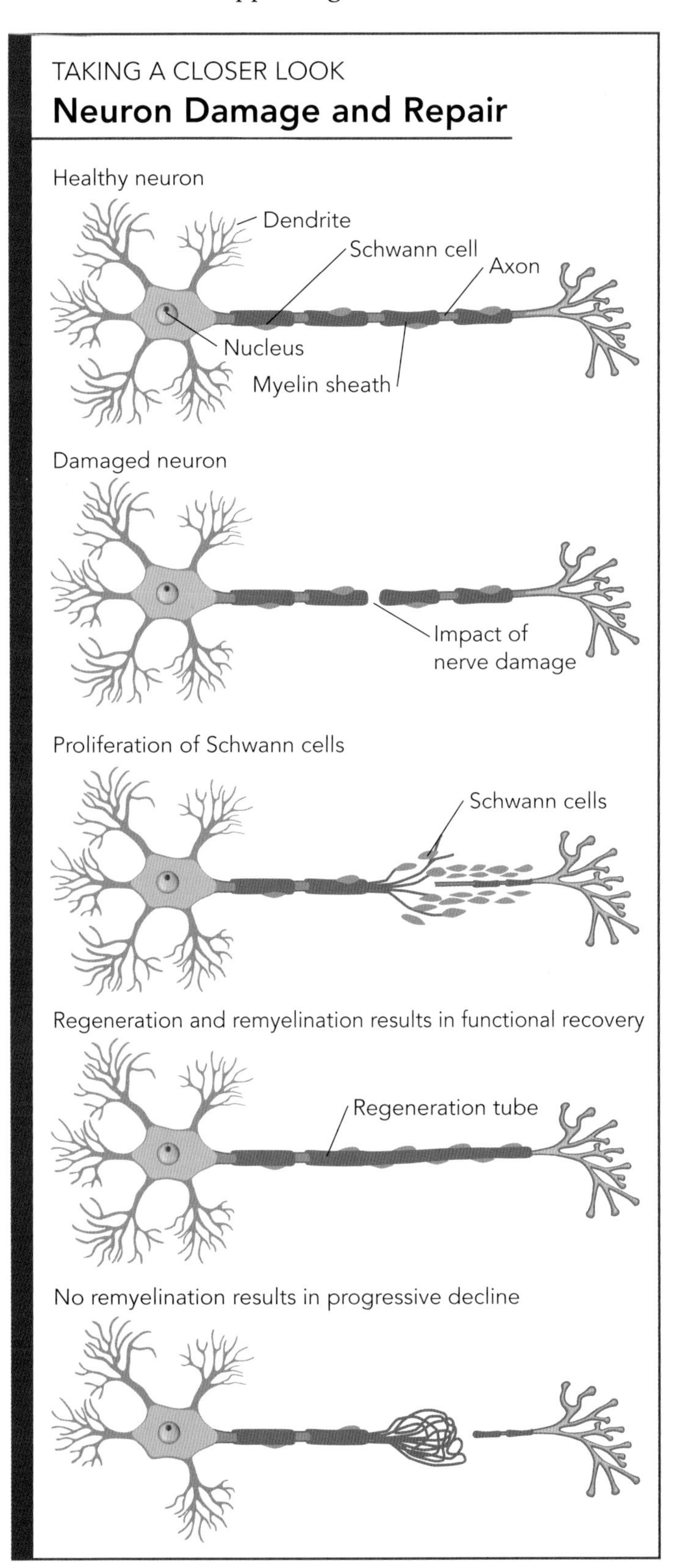

When a nerve is badly damaged, proteins and other vital substances produced in the neuron cell bodies cannot be transported all the way out to the ends of their axons. The distal (farther away) portions of the axons—the part beyond the injury—begin to break down without these nutrients. This is known as Wallerian degeneration. However, the Schwann cells near the injured area multiply and begin to form a protective tube. This "tube" helps align the damaged ends of the axons as they regenerate. Further, the Schwann cells secrete growth factors to promote axon regeneration. Therefore, nerve damage in the PNS does not always result in permanent loss of function.

It is a different story in the CNS. Recall that myelination in the CNS is due to the presence of oligodendrocytes. Unlike the Schwann cells in the PNS, oligodendrocytes do not have the capability of supporting regeneration of a damaged axon. For this reason, damage to the brain or spinal cord is more serious and more likely to be permanent than peripheral nerve injury.

SO SIMPLE YET SO COMPLEX — Designed by the Master

The foundation of our thinking in every area of our lives should be the Word of God. How we understand the world, how we approach our daily tasks, how we view and treat our fellow man — these things should be based on the principles we find in the Bible.

Unfortunately, too many people are so strongly influenced by the views of the world that they reject the direct teaching found in God's Word. These people view the world around them as just a chemical accident. Matter somehow just came into existence all on its own billions of years ago. Then everything in our world just created itself. Millions of years of chemicals banging together resulted in something as incredibly complex as the human body.

Even though we've only just begun our study of the nervous system, I'll bet you are already getting the idea of how complex just this one body system truly is. Do you really think it could have just created itself, all on its own? No, neither do I.

In the Book of Genesis, we are told

> In the beginning God created the heavens and the earth (Gen. 1:1).

There is an all-powerful God who indeed created all things. The earth, the living creatures, the sun and moon, the planets, the stars in the sky—these things did not come into being as the result of an accident. They are not the result of time and chance. They are the work of our wonderful Creator.

Even more, you and I are not the products of chance. We are special creations.

> Then God said, "Let Us make man in Our image, according to Our likeness" (Gen. 1:26a).

As we continue our study of the human body, we need to always remember that the complex systems we study bear the unmistakable mark of the Master Designer. The enormous complexity of the body should remind us constantly of God's wisdom and creativity. We should also be reminded of His boundless love for us that He should take such care in our creation.

NERVE SIGNALS

A fundamental characteristic of nervous tissue is that it is excitable. That is, it is capable of being stimulated to produce an electrical signal. The stimulus is some sort of change in the environment, a change in temperature for a temperature-sensitive nerve cell in the skin, or a change in light for the light-sensitive cells in the eye, or a change in the neurotransmitters around the dendrites of sensory neuron. The stimulus triggers an electrical signal that then travels down the nerve cell membrane.

We will now begin to explore this amazing phenomenon in depth.

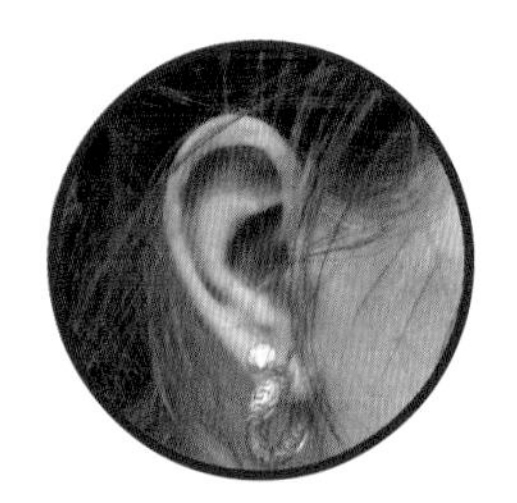
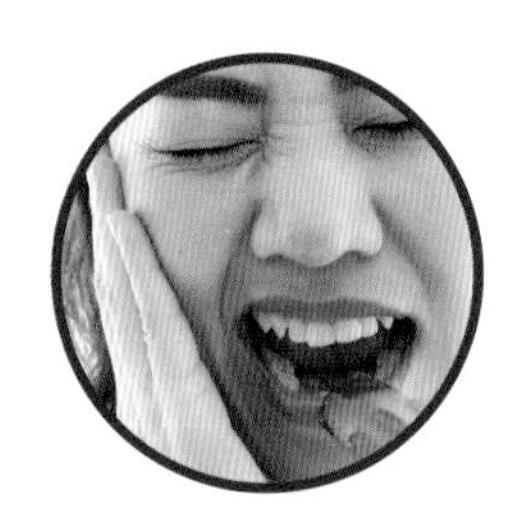

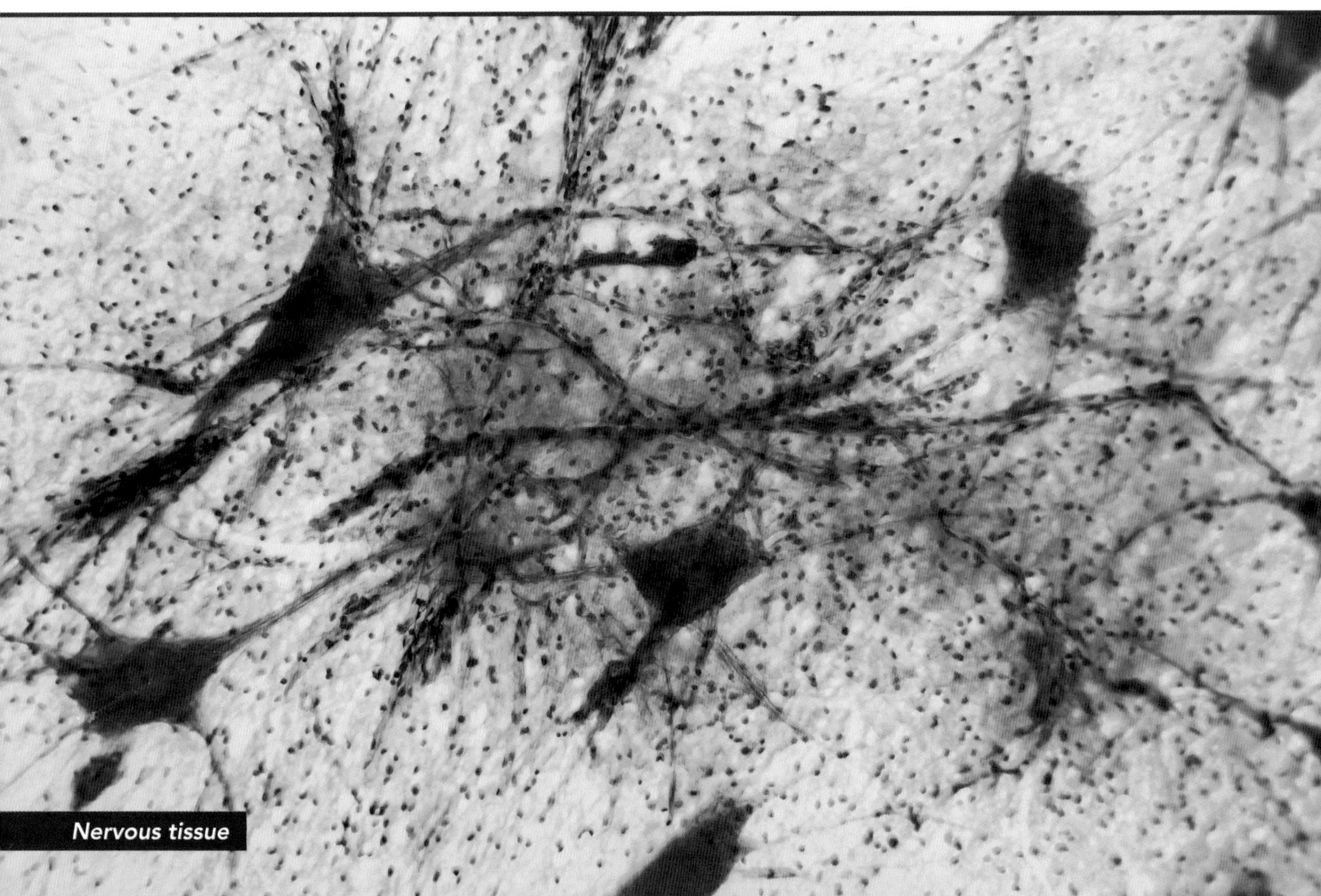

Nervous tissue

Basic Principles

For a neuron to function properly, it must be able to produce an electrical impulse and then transmit it along the length of its cell membrane. Production of this traveling electrical impulse depends on the movement of charged particles into and out of the cell. As charged particles—called ions—move across sequential parts of the neuron cell membrane, the electrical impulse moves along the length of the neuron. Let's see how these electrical impulses are generated and how they travel.

Your body itself is electrically neutral. This means that the number of negative particles equals the number of positive particles (or almost equals, you know, give or take a particle or two...). The extracellular fluid (fluid outside cells) is electrically neutral, as is the intracellular fluid (the fluid inside cells). So if everything is neutral, where do these traveling electrical impulses come from? These electrical impulses are created by the movement of ions across the cell membrane of the neuron.

TAKING A CLOSER LOOK

Concentration Gradient

The cell membrane (also called the plasma membrane) of a neuron contains special proteins that extend through the full thickness on the membrane. The special proteins have channels or pores that allow passage of certain ions (charged particles) into and out of the neuron. These channels open and close in response to certain stimuli. For example, some channels open when in the presence of specific chemicals (perhaps a neurotransmitter, as we will see later). Other channels ope
a change in voltage across the membı

When these channels open, ions move through them based on certain chemical principles. First, ions move from areas of higher concentration to areas of lower concentration. For example, let's say that in a certain situation there was a high number of potassium ions in the cytosol of the cell and a much lower number of potassium ions in the extracellular fluid around the cell. Then, a potassium (K^+) channel opens in the cell membrane that allows movement of K^+ ions. What do you think will happen? You are correct. K^+ from the inside of the cell will rush out of the cell toward the area of lower concentration. This is called a concentration gradient.

Second, ions move toward regions of opposite charge. Positively charged ions move towards regions of negative charge, and negatively charged ions move towards areas of positive charge. This is called an electrical gradient.

The Resting Membrane Potential

As previously mentioned, the environments inside and outside the cell are each electrically neutral. That is, there are equal numbers of positive and negative charges. However, there does exist a small electrical difference across the cell membrane. This electrical difference is called the resting membrane potential.

If you measure the electrical difference between the area outside the cell membrane and the cytosol inside, you will find a difference of about minus 70 millivolts (- 70mV). The inside of the cell is negatively charged when compared with the outside of the cell. When a membrane potential, that is, a charge across a membrane, exists, the membrane is said to be polarized.

Membrane polarization is possible because of the different concentrations of ions in the intracellular and extracellular fluids and their abilities to cross the cell membrane. There is a much higher concentration of potassium ions (K^+) inside the cell than in the extracellular fluid surrounding it. And there is a higher concentration of sodium ions (Na^+) in the extracellular fluid than there is inside the cell. Inside the cell, there is electric neutrality because there are enough negative ions to balance the positive ones. And in the extracellular fluid outside the cell, there is electric neutrality for the same reason. But across the cell membrane, there is a measurable electrical difference, the resting membrane potential.

How is this possible? Remember we said K^+ and Na^+ ions are able to travel through the cell membrane. But the journey through the membrane is easier for some ions than others. Potassium and sodium ions differ in their ability to cross the cell membrane. At rest, the cell membrane is much more permeable to K^+ ions than to Na^+ ions. In other words, the protein channels that allow ions to pass are much more open to the movement of K^+ than to Na^+.

If there is a higher concentration of K^+ inside the cell than outside, in what direction do K^+ ions move? If you said, "Potassium ions move out of the cell," you would be right. By contrast, Na^+ is more concentrated outside the cell. Therefore, Na^+ "wants" to move into the cell. But since K^+ moves through the cell membrane much more easily than Na^+, lots of K^+ flows out, but much less Na^+ gets in. Therefore, on balance, more positive charges move out of the cell than move in. This leaves the inside of the cell a little bit negative compared to the outside. Positive ions build up outside the cell, leaving a relatively negative charge on the inside. This electrical difference is about -70mV.

TAKING A CLOSER LOOK

Establishing Ion Gradients

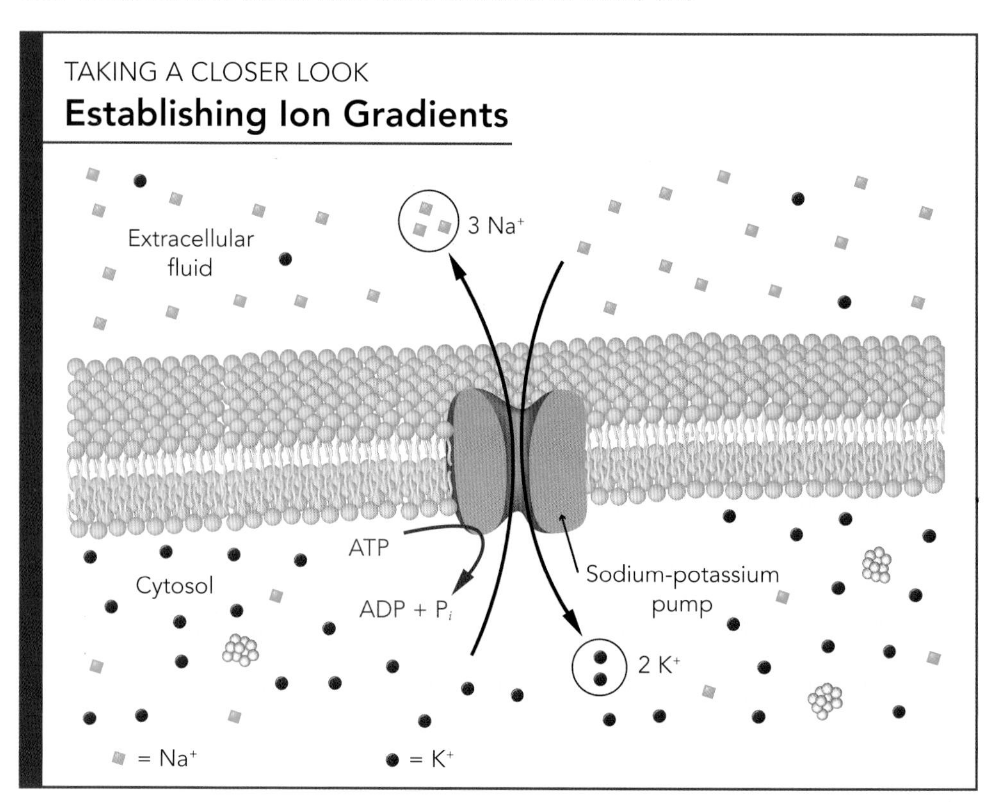

TAKING A CLOSER LOOK

Nerve Impulse Action Potential

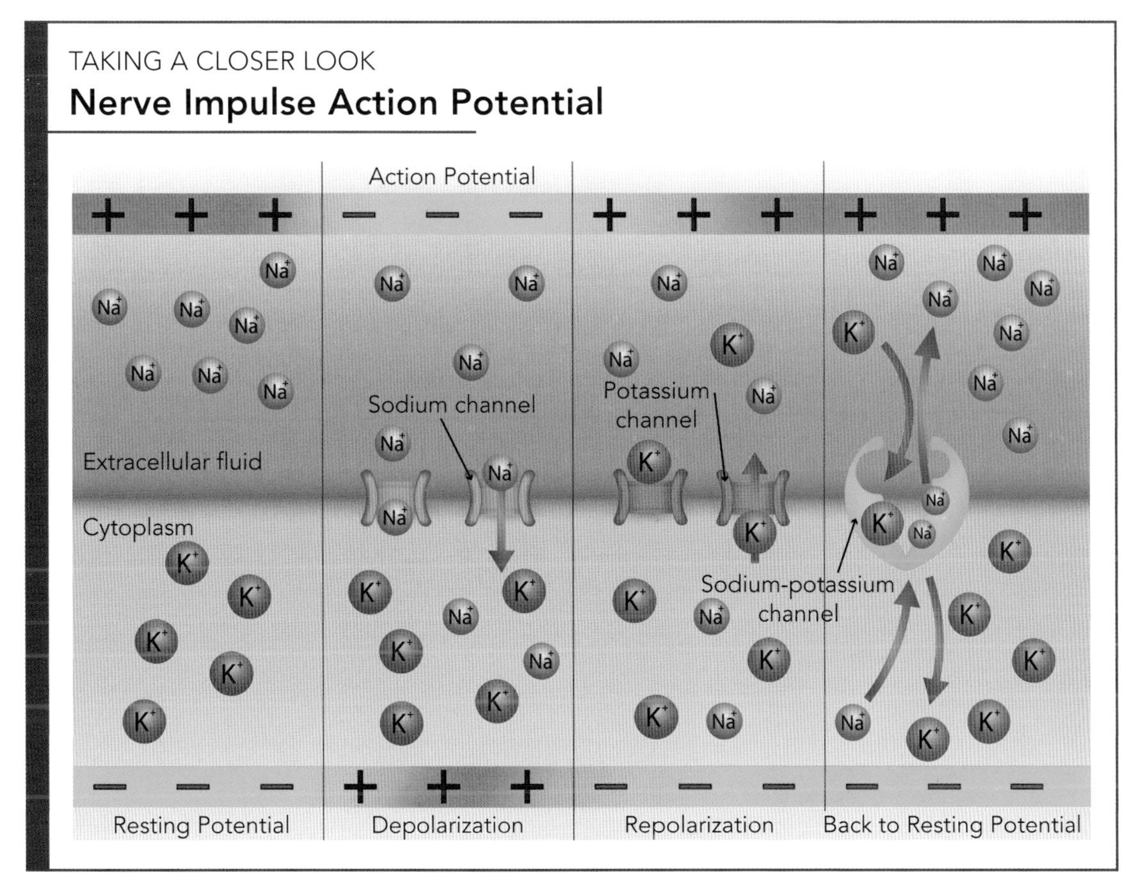

But wait a minute. If you think about it, eventually the concentration of K^+ and Na^+ would be the same on both sides of the membrane, right? After all, these ions move from higher concentration to lower concentration. Sooner or later they would reach a state where the concentrations are equal on both sides of the membrane. So would this not eliminate this electrical potential across the membrane? Yes, it would....but (isn't there always a "but"?).

There is a special type of pump in the cell membrane that helps keep the resting potential stable. This is called a sodium-potassium pump. The sodium-potassium pump transports Na^+ out of the cell and K^+ into the cell. This process maintains the concentration differences that create the resting membrane potential.

The sodium-potassium pump works hard, and it requires energy to do its job. Energy to fuel this sort of work is supplied by a special molecule called adenosine triphosphate (ATP). ATP stores chemical energy and is ready to supply it for all sorts of cellular work. ATP is the main energy currency of the cell.

The sodium-potassium pump is perfectly designed for its job. It is obviously meant to maintain the resting membrane potential because the number of Na^+ and K^+ it transports is not equal. For every two K^+ brought into the cell, three Na^+ are pumped out. This means that this pump maintains the -70mV inside the cell, because it consistently pumps more positive ions out of the cell than it brings in!

(You've probably noticed by now that we describe the membrane potential from the "point of view" of the inside of the cell. When there are more positive ions outside the cell than there are inside, the inside is negative compared to the outside.)

The Action Potential

We've talked about how the neuron's cell membrane gets and stays polarized. The sodium-potassium pump pushes more positive charges to the outside than it brings in. A polarized membrane—an electrical difference between the inside and the outside—must exist before a neuron can send an electrical signal. So how exactly does a neuron send an electrical signal?

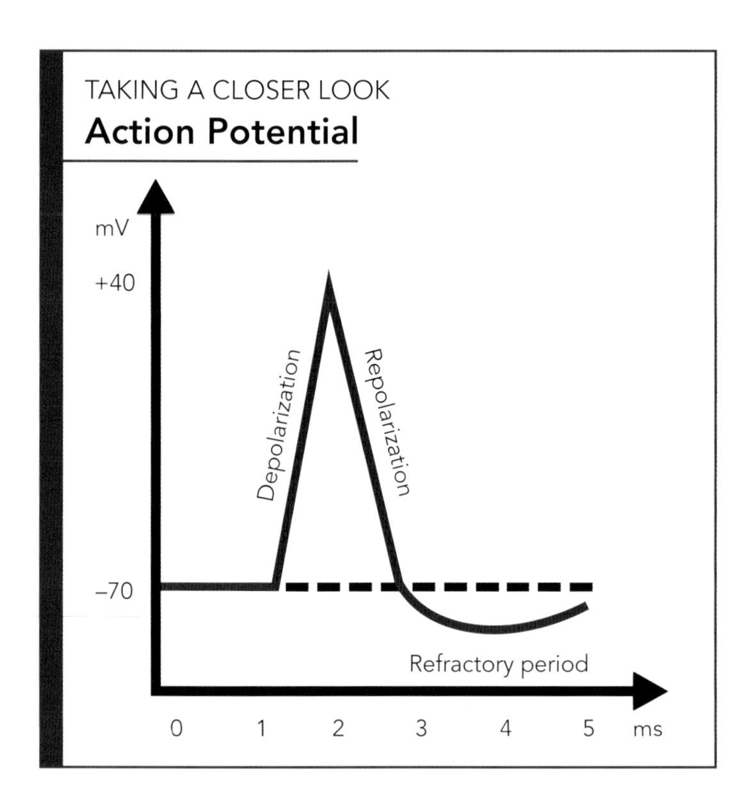

Action potentials occur only in axons. When an axon is presented with an adequate stimulus, the nerve impulse is triggered. However, not every stimulus is strong enough to trigger the action potential. You see, the membrane potential must reach a certain level of depolarization (that is, it must become sufficiently less negative than its usual -70mV) to initiate the action potential. In a typical neuron, this level is -50mV or so. This is called threshold level. Any depolarization not reaching this level is sub-threshold and will not trigger a nerve impulse. A very important thing to remember about an action potential is that it is an all-or-none event. When a stimulus is received, there is either a full action potential or there is no action potential at all.

The electrical signal a neuron generates is called a nerve impulse. It is also called an action potential. An action potential is a change in the membrane potential from negative (more positive charges outside) to positive (more positive charges inside) and then back again.

The first thing that happens during an action potential is called depolarization. During depolarization, the membrane potential becomes less and less negative, and then positive. In other words, once a membrane is depolarized, there are more positive charges on the inside than there are outside. (Remember, these "negatives" and "positives" represent the point of view from the inside of the nerve cell.)

The next step is called repolarization. During repolarization, the membrane potential is reset, becoming negative once more. Once it is repolarized, there are more positive charges outside than inside. Having had its resting membrane potential reset, the membrane is prepared for the next action potential.

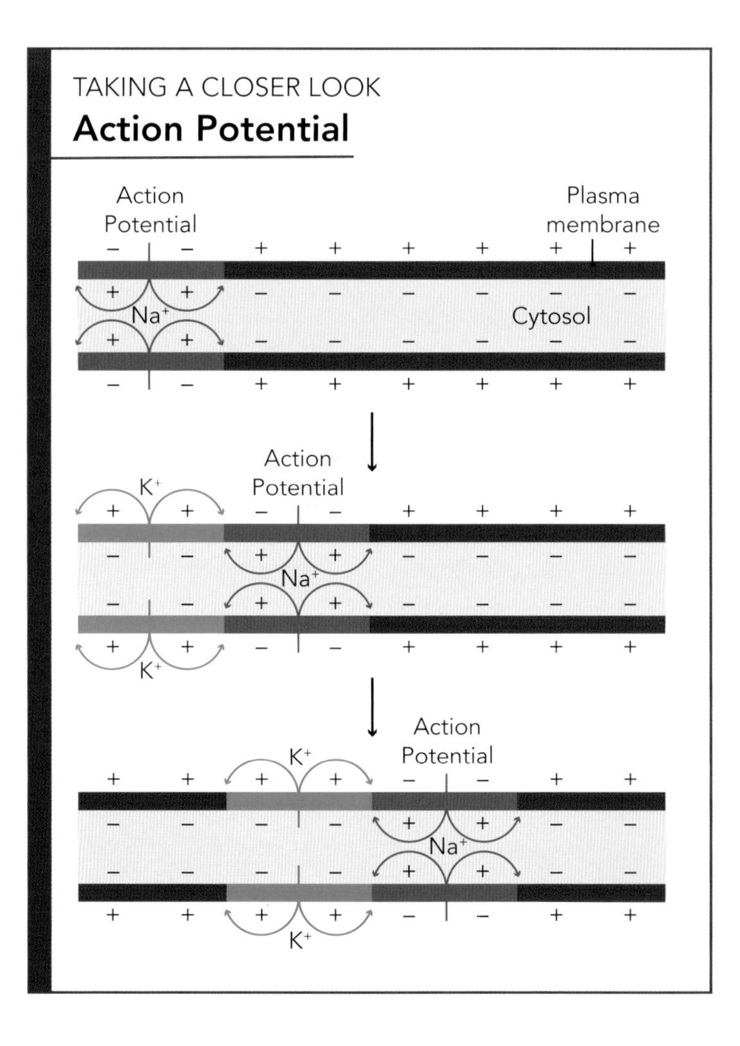

Depolarization

At rest the cell membrane is permeable to both K^+ and Na^+. That means both K^+ and Na^+ ions can pass through the membrane. This permeability is possible because of special proteins that create a channel to facilitate the movements of these ions through the cell membrane. There are, however, other special channel proteins involved with the movement of ions. These are called voltage-gated channels. These channels are opened when voltage across the membrane changes. As we will see in a moment, there are voltage-gated channels for Na^+, and there are other voltage-gated channels for K^+.

Resting Potential: At the resting potential, voltage-gated Na^+ channels and voltage-gated K^+ channels are closed. The Na^+/K^+ pumps moves K^+ ions into the cell and Na^+ ions out of the cell.

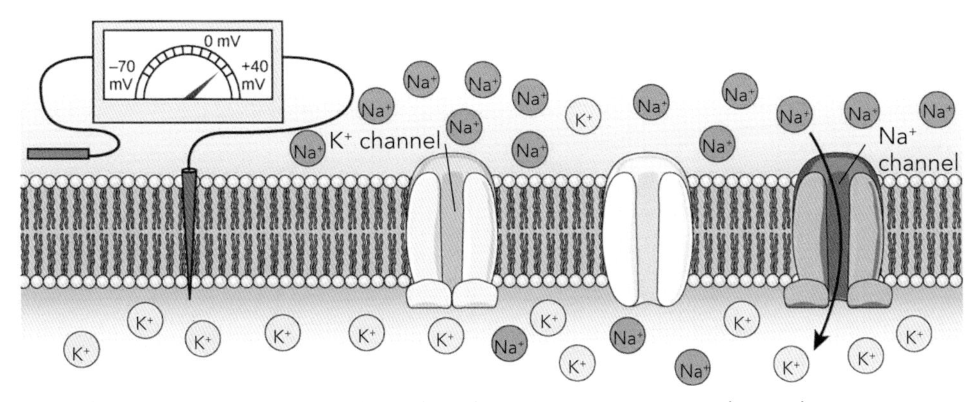

Depolarization: In response to a depolarization, some Na^+ channels open, allowing Na^+ ions to enter the cell. The membrane starts to depolarize (the charge across the membrane lessens). If the threshold of excitation is reached, all the Na^+ channels open.

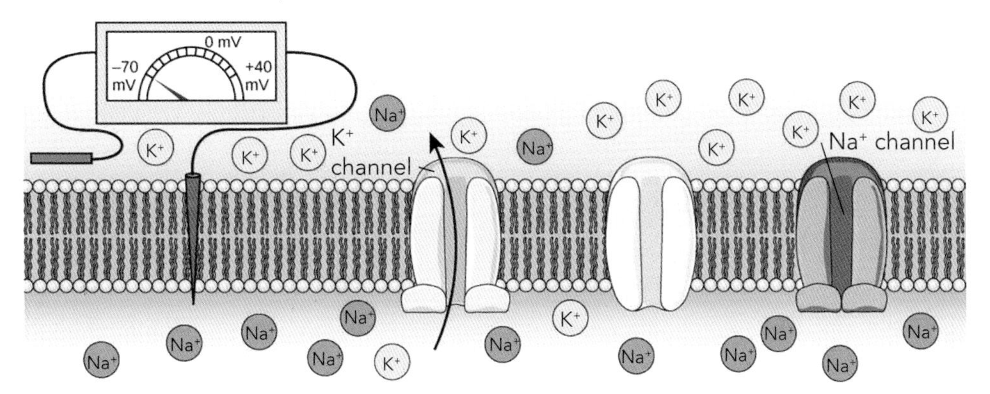

Hyperpolarization: At the peak action potential, Na^+ channels close while K^+ channels open. K^+ leaves the cell, and the membrane eventually becomes repolarized.

When a stimulus reaches the voltage-gated Na^+ channels, the channels open and Na^+ pours into the cell. This influx of positive ions makes the inside of the cell less negative. If the threshold level of -50mV is reached, many more voltage-gated Na^+ channels open, and much more Na^+ flows into the cell. At this point depolarization is in full swing, and the action potential moves on to a neighboring portion of the membrane, continuing down the entire length of the axon. At the end of the depolarization phase the membrane potential is around +30mV. You see, the inside of the cell is now positive with respect to the outside because a flood of Na^+ ions has resulted in more positive charges inside the cell than outside.

This remarkably complex and precise series of events happens in less than a thousandth of a second.

Repolarization

At the end of the depolarization phase, the membrane potential is around +30mV. Obviously this is far from the level of the resting membrane potential. In this state the neuron is not able to trigger another action potential. The neuron's negative resting membrane potential must be reset before another action potential can travel along that portion of the axon.

When the membrane potential reaches +30mV, the voltage-gated Na^+ channels begin to close. Therefore, movement of Na^+ into the cell stops almost completely. Then voltage-gated K^+ channels open.

K^+ rushes out of the cell through these channels as soon as they open. Remember that the resting concentration of K^+ is higher inside the cell than outside. When the "door" is opened, these ions rapidly move from an area of high concentration to an area of low concentration. Furthermore, the depolarization left the outside of the cell more negative than the inside. Positive charges are attracted to a more negatively charged area. Potassium ions therefore flood out of the cell through their channels.

As the membrane potential returns to negative—in other words, when there are more positive charges outside the cell than inside—these voltage-gated channels close, and the permeability of the membrane returns to its normal resting state. In some instances the outflow of K^+ is so rapid during repolarization that the membrane potential will overshoot to near -90mV. This is called hyper-polarization. When the voltage-gated K^+ channels are all closed, normal membrane potential is rapidly restored.

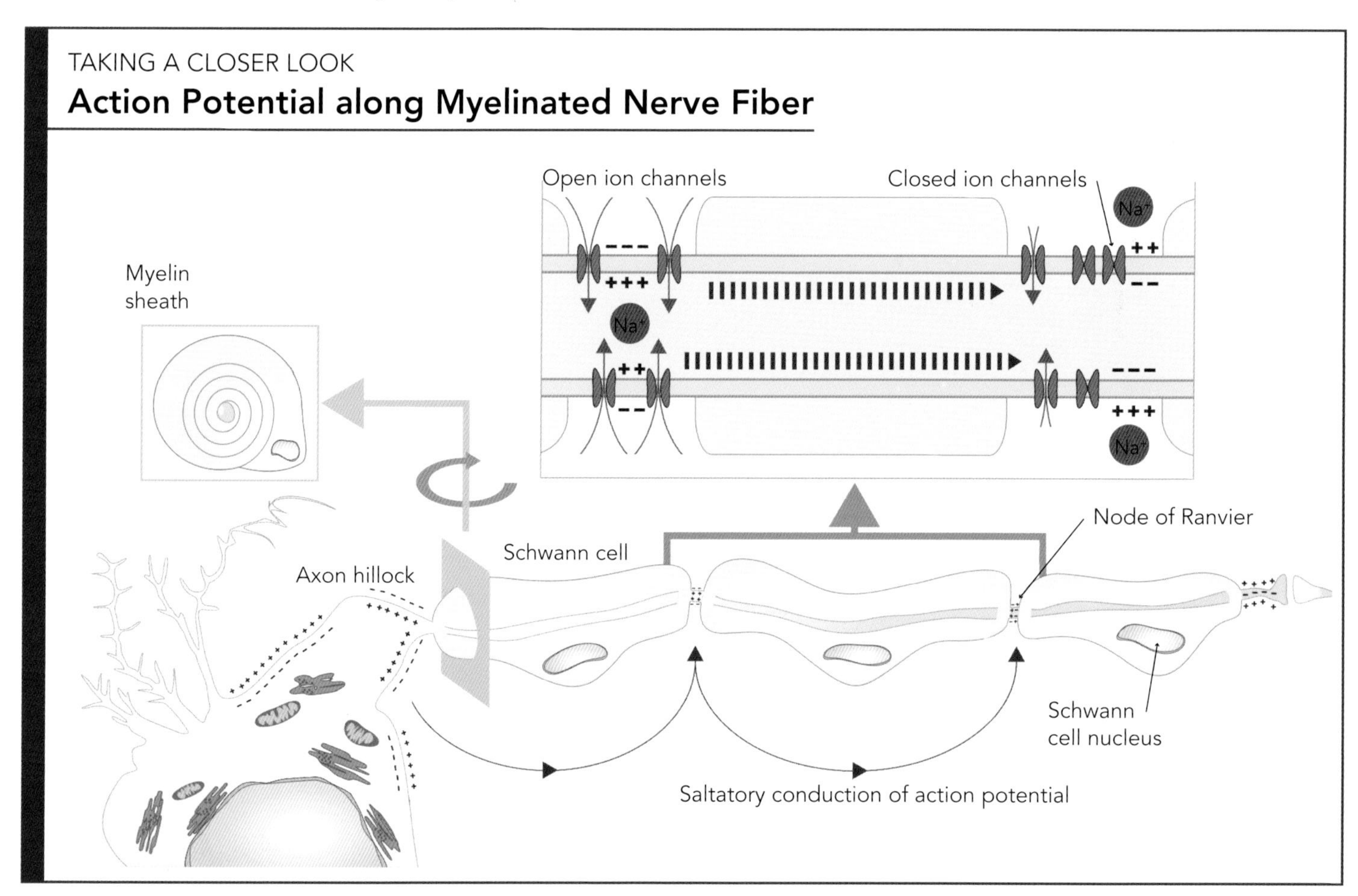

Repolarization, which prepares the axon for the next action potential, takes only a few thousandths of a second.

Re-establishing Normal Resting Ion Concentrations

Even though a membrane is repolarized, or even hyperpolarized, after the potassium ions rush out of the cell, the sodium and potassium ions are not in their normal resting places. Normally, the K^+ concentration inside is greater than outside, and the Na^+ concentration outside is greater than inside. How do the ions return to this normal state?

Remember the sodium-potassium pump? The sodium-potassium pump gets busy. Using energy from ATP, the pump pushes out three Na^+ for every two K^+ that it brings into the cell. Soon the normal resting state is reestablished.

Conduction of Action Potentials

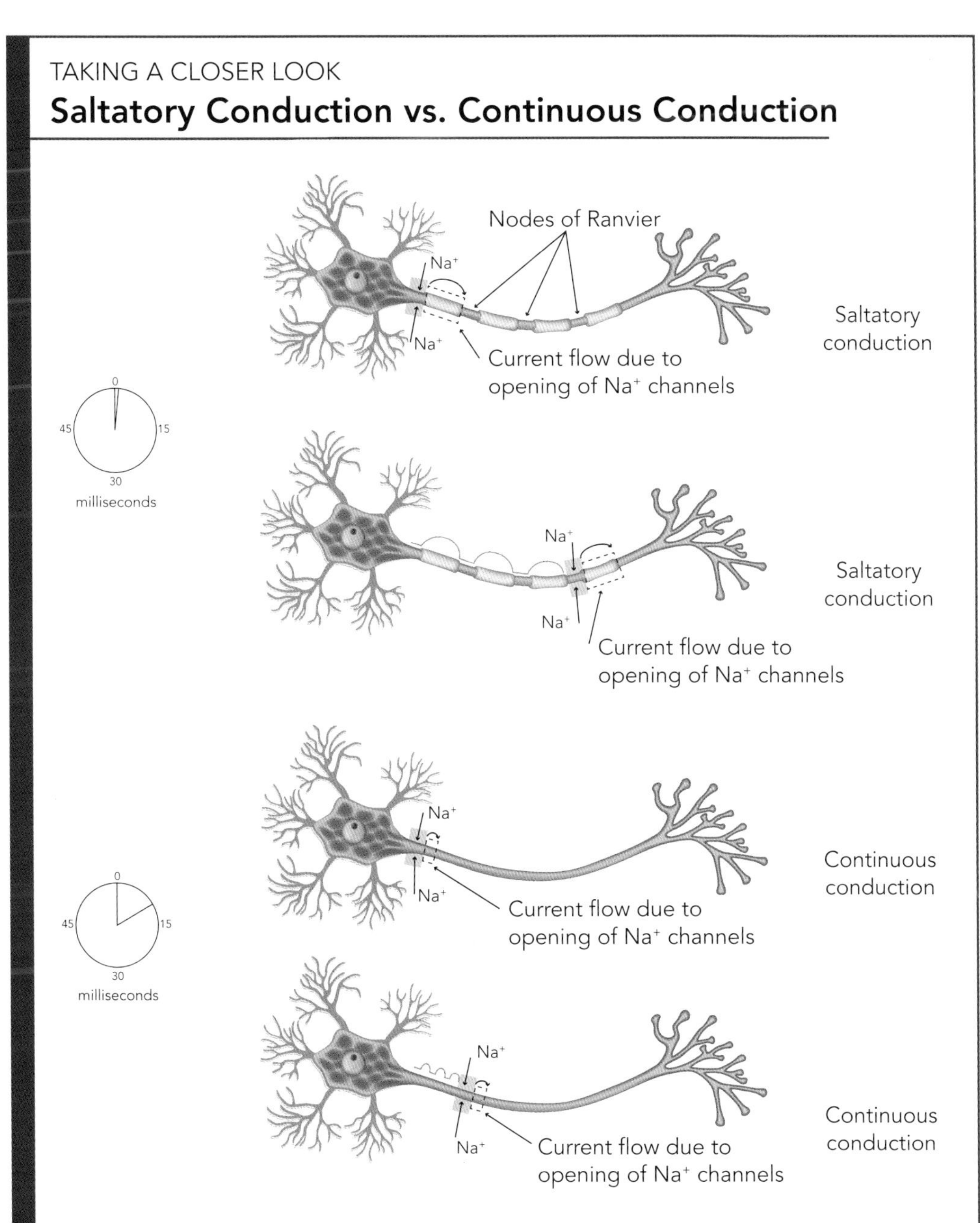

An action potential moves down the length of the axon membrane like a series of dominoes each knocking the next down. A voltage change triggers voltage-gated channels to open in one area, producing the action potential there. This voltage change triggers the same response in the neighboring area of the membrane. Here one region directly triggers the next, and the next, and the next, and so on. This is known as continuous conduction. Continuous conduction occurs in unmyelinated axons. Because a bit of time is required to repolarize each section of the axon membrane before another action potential can be created there, continuous conduction is slow compared to the sort of conduction that occurs in myelinated axons.

In myelinated axons, nerve impulse conduction is much more rapid. This is due to the myelin covering the axon itself. The myelin covering insulates the axon. Further, in myelinated axons, voltage-gated channels are concentrated near the gaps between the Schwann cells, the nodes of Ranvier. It is not necessary to repolarize then entire surface of the axon, just the portions at the nodes of Ranvier. By generating local currents around the myelin sheath, the action potential seems to "leap" from one gap to the next. The myelin sheath therefore speeds up conduction of the action potentials through the length of the axon. This is known as saltatory (from saltare, meaning "to leap or hop") conduction.

"How much faster?" you are probably wondering. Well, nerve impulse conduction speed varies based on the thickness of the axon and how much myelin covers it. But to get an idea of the sort of speeds we are talking about, figure a small unmyelinated axon might transmit nerve impulses at around 2 miles per hour. Depending on its diameter and degree of myelination, a myelinated axon might transmit nerve impulses at 30 to 300 miles per hour. When you realize how dramatically myelination speeds up nerve impulse conduction, you can understand how debilitating a disease that destroys myelin must be.

Strong Versus Weak Signals

So we have seen that action potentials are all-or-nothing phenomena. When triggered the action potential fires completely. There is no such thing as a partial action potential. In a system as finely regulated as the nervous system, isn't this a rather clumsy arrangement? After all, if a stimulus triggers an action potential, and one action potential is pretty much like all others, how can such fine control be achieved?

We need to understand the difference between how a nerve responds to a weak stimulus compared to its response to a strong stimulus.

It is obvious that there are varying degrees of sensations we perceive every day, right? Water can be cold, warm, or hot. We can feel a gentle breeze on our face or strong winds on a stormy day. Walking on the beach, we can feel that the sand under our feet is warm and soothing or very hot and unpleasant. How do our neurons help us tell the difference?

Because all action potentials are the same, a "warm" stimulus triggers the same action potential as a "hot" one. However, a stronger stimulus triggers more action potentials in a given time than a mild one. A stimulus from sensory receptors sensing a strong wind will trigger more frequent action potentials by comparison to those triggered a mild breeze.

Also, as a general rule, more intense stimuli tend to activate more sensory receptors than mild stimuli do. Therefore, intense stimuli trigger a flurry of action potentials from more receptors than mild stimuli.

If you step barefoot onto hot sand, more action potentials are triggered in more temperature receptors than when you burrow your toes into soft, warm sand that is just right. Therefore, your CNS receives more input from a greater number of

neurons from your foot stepping on the hot sand than from your toes curling in the warm sand.

Graded Potentials

Since action potentials occur only in axons, it is reasonable to ask how a nerve impulse is transmitted through the rest of the neuron. Nerve impulses travel through dendrites and neuron cell bodies using graded potentials.

When a dendrite receives a stimulus, there is a small change in the membrane potential in that area. However, an action potential is not triggered. Instead, a graded potential is generated. The stimulus causes a small number of channel proteins to open. This results in some ion movement across the cell membrane. However this small membrane response travels only a short distance and then fades away.

While an action potential is all-or-none, a graded potential varies in degree. Graded potentials vary with the strength of the stimulus. The greater the stimulus, the greater number of ion channels open. And the stronger the graded potential, the farther it travels. If this type of impulse travels far enough, reaching the initial part of the axon, it can ultimately lead to a full action potential.

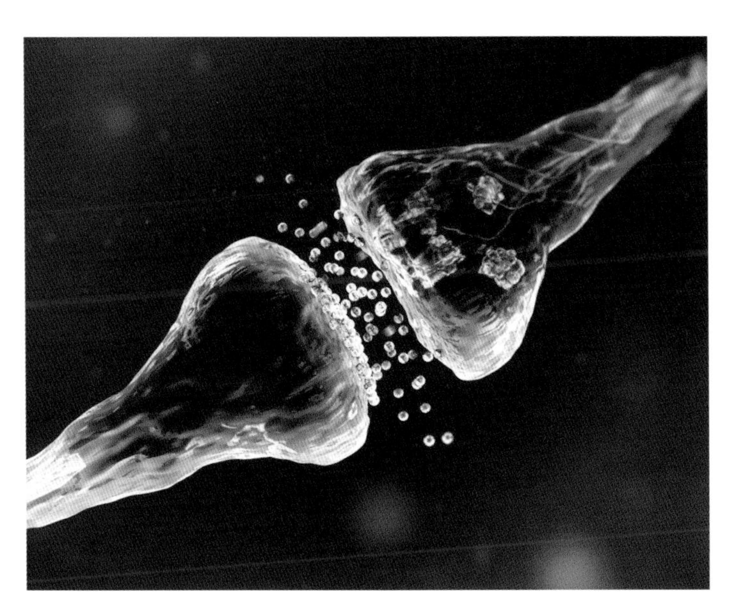

The Synapse

We have seen how the neuron transmits a signal along its length. A stimulus triggers graded potentials in the dendrites, and if a sufficiently strong impulse reaches the initial part of the axon, an action potential is triggered and passed down the axon's length. Thus we've now seen how a nerve impulse travels the whole length of a neuron.

That's all well and good you might say. But when a nerve impulse reaches the end of the axon, how does that impulse get to the next nerve, or to a muscle, or to anywhere for that matter?

A nerve signal is passed along via something called a synapse. A synapse is the place where a neuron communicates with another neuron or with a muscle cell.

At the far end of an axon are small branches called axon terminals. Axon terminals are positioned very close to the cell membrane of the next cell, the cell that is to receive the signal. The axon terminal, the membrane of the next cell, and the space between them (known as the synaptic cleft), are collectively known as a synapse.

The neuron that carries the impulse towards the next cell is called the presynaptic neuron (pre- because this nerve is located "before" the synapse). The neuron that is to receive the nerve impulse is called the postsynaptic neuron (post or "after" the synapse). Neurons are usually not just pre- or post-synaptic. Most neurons are both. A given neuron is postsynaptic at its dendritic end where it receives impulses. The same neuron is presynaptic at its axon terminals where it sends impulses to the next cell. These terms are useful, however, when we describe the neurons around a particular synapse.

Synapse and neuron cells sending chemical signals.

There are two basic types of synapses: electrical and chemical.

At an electrical synapse, the action potential is transmitted directly to the next cell. This occurs by means of special connections called gap junctions. Gap junctions connect the cytosol of adjacent cells. Gap junctions allow the nerve impulse, like an electrical current, to pass directly to the next cell.

The most common type of synapse is the chemical synapse. In a chemical synapse, the nerve impulse does not travel directly from axon terminal to the post-synaptic neuron but must send a different sort of signal, a chemical signal, across the synaptic cleft. Let's take a closer look.

Chemical Synapse

A chemical synapse is designed to transfer nerve signals by releasing special chemicals called neurotransmitters. This synapse is an incredible design.

The axon terminal of the presynaptic neuron is positioned adjacent to receptor region on the post-synaptic neuron. The space between the two cells is extremely small. It is usually between 20 and 40 nanometers. That is 20 to 40 billionths of an inch! (Not much wiggle room there.)

The axon terminal of the presynaptic neuron contains many small sacs, called synaptic vesicles. These vesicles contain the neurotransmitters. When an action potential reaches the axon terminal, voltage-gated Na^+ channels open, and Na^+ rushes into the cell. This is the same process we covered earlier. However, in the axon terminal there are also voltage-gated Ca^{2+} (calcium ion) channels. The action potential triggers these also, and Ca^{2+} pours into the cell. This increase in Ca^{2+} inside the cell is a signal for the cell to begin to release the neurotransmitters.

Neurotransmitters are released from the synaptic vesicles into the synaptic cleft by exocytosis. This means that the vesicles containing the neurotransmitters merge with the cell membrane and release their chemical messengers into the synaptic cleft.

A single action potential will result in the release of only a small amount of neurotransmitter. If multiple, frequent action potentials are received, then more neurotransmitter is released. Thus a greater stimulus causes a larger of amount of neurotransmitter to be released into the synaptic cleft.

TAKING A CLOSER LOOK

Synapses Can Occur in Many Locations

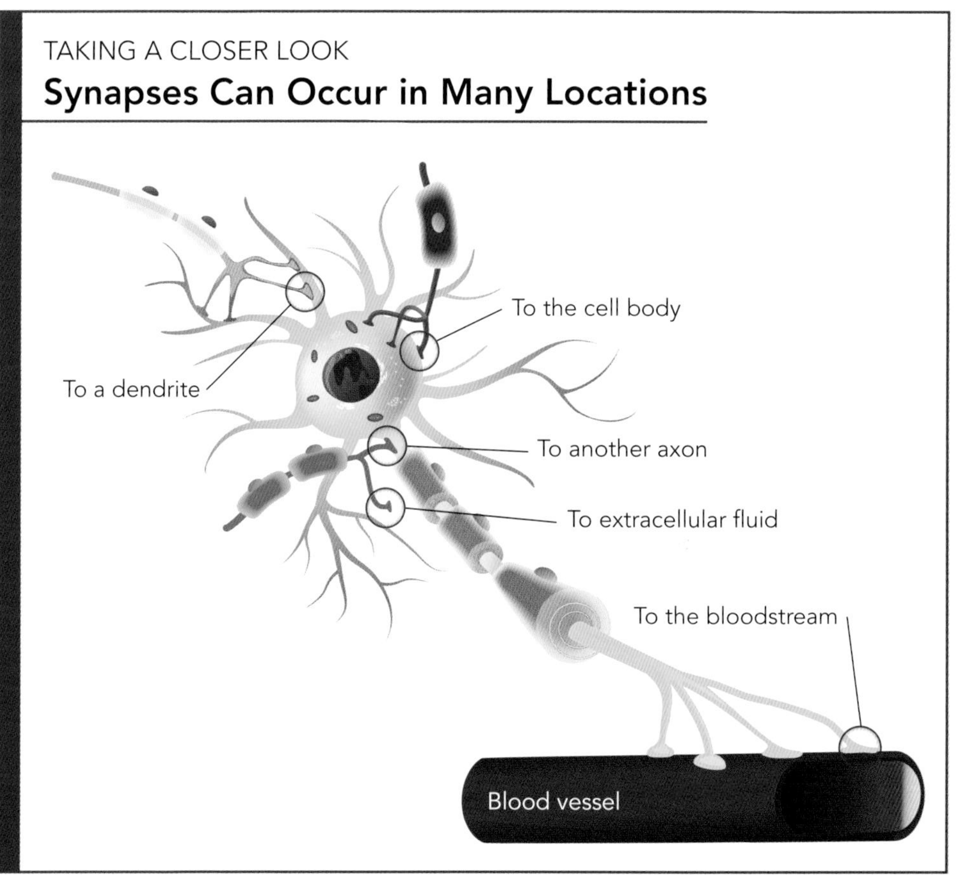

The neurotransmitter diffuses across the synaptic cleft and binds to special receptor proteins on the postsynaptic neuron. These receptor proteins then allow movement of ions across the cell membrane of the postsynaptic neuron. These receptor proteins are called chemically-gated channels. This means that they open and close not because of a change in voltage but rather due to the presence of certain chemicals, the neurotransmitters.

Now comes the really cool part. Neurotransmitters can either excite or inhibit the postsynaptic neuron. Some neurotransmitters cause the postsynaptic cell membrane to slightly depolarize. These promote excitation of the post-synaptic neuron. Other neurotransmitters cause the post-synaptic cell membrane to become slightly hyper-polarized, which tends to inhibit the production of another nerve impulse.

Remember that when the inside of the cell becomes less negative, and thus is brought closed to its threshold voltage, it is being depolarized. If enough excitatory neurotransmitter is released across the synaptic cleft, then a graded potential strong enough to trigger an action potential may result. Excitatory signals bring that postsynaptic neuron closer to firing an action potential.

The opposite is true if the neurotransmitter causes the postsynaptic neuron to become more negative inside. This takes the neuron farther away from the threshold. This is hyper-polarization.

Man, that's a lot to take in, so if you're feeling overwhelmed, you are probably normal. It can get quite complicated. (Remember this sentence when you are in medical school in a few years.) It really comes down to something simple. Let's have a look

TAKING A CLOSER LOOK

The Synapse

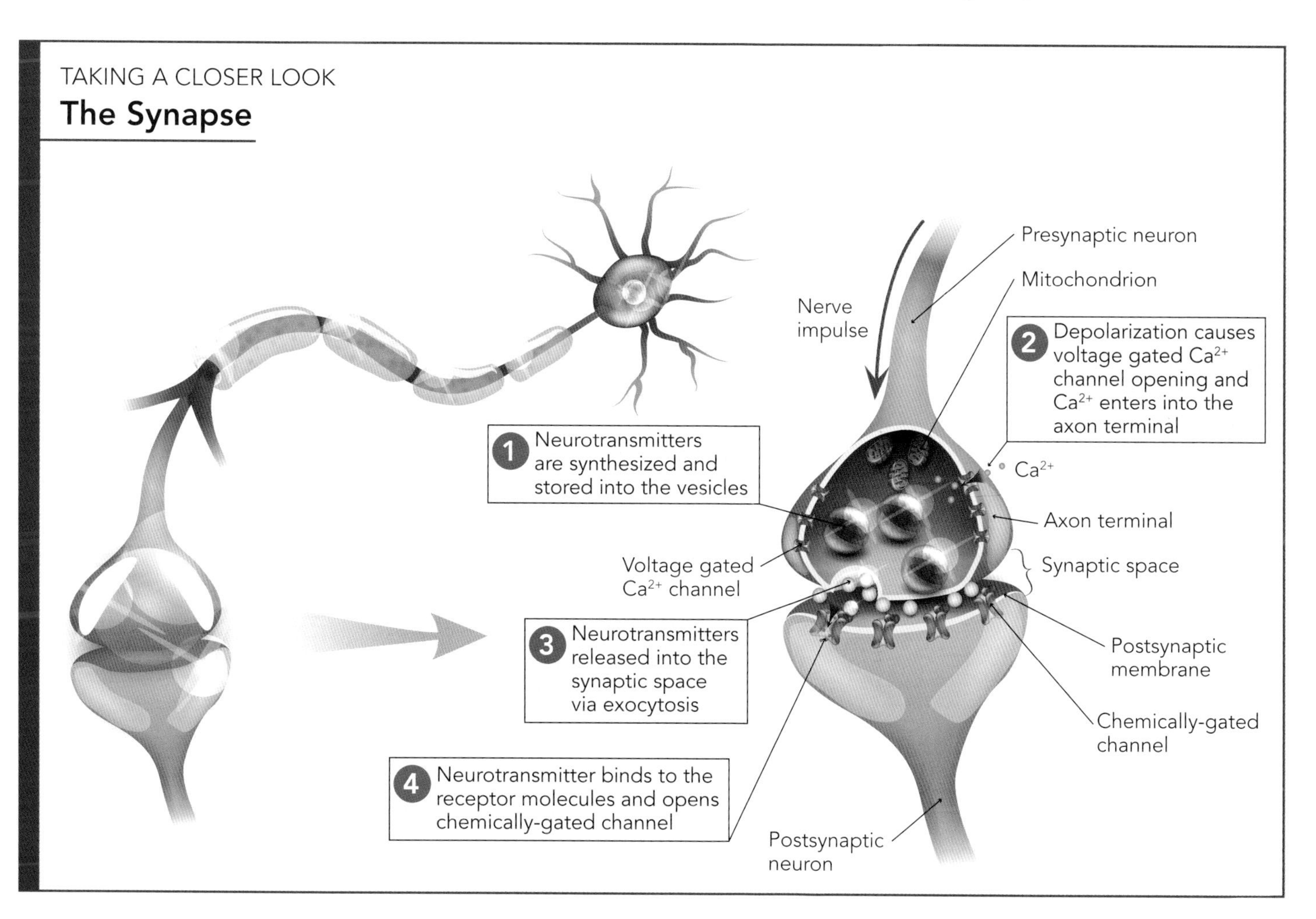

at the big picture, and I think this will make a lot of sense to you.

The Role of the Synapse

The synapse is simply a mediator. It sort of "takes a vote." It allows many different stimuli or inputs to "have their say" in determining whether the postsynaptic neuron fires or not. See, easy.

Because the synapse works the way it does, it prevents any one single stimulus from triggering the postsynaptic neuron. It takes a significant stimulus to fire that neuron. The presence of both excitatory and inhibitory neurotransmitters makes it possible to achieve very fine control of neuron firing. This is crucial for the proper functioning of the nervous system. For example, if a postsynaptic membrane receives four excitatory stimuli and four inhibitory stimuli, the signals basically cancel each other out (4-4=0), so the postsynaptic neuron does not fire. If there are 10 excitatory stimuli and only two inhibitory stimuli, the postsynaptic membrane will be brought close to threshold, perhaps enough to trigger the action potential. You can imagine that if every stimulus, no matter how mild, were able to elicit a big response, your nervous system would be rather overloaded, and you would overreact to everything.

Therefore, unlike electrical synapses that directly transmit signals to adjacent cells, chemical synapses do not transmit electrical signals at all. They allow electrical signals in the presynaptic neuron to be converted to chemical signals. These chemical signals are then converted back into electrical signals in the postsynaptic neuron. While at first glance this might seem overly complex, it does allow very fine control of nerve signal transmission.

You really should not try to make it any harder than that. The synapse is a magnificent design from the Master Designer!

TAKING A CLOSER LOOK

Neurotransmitter

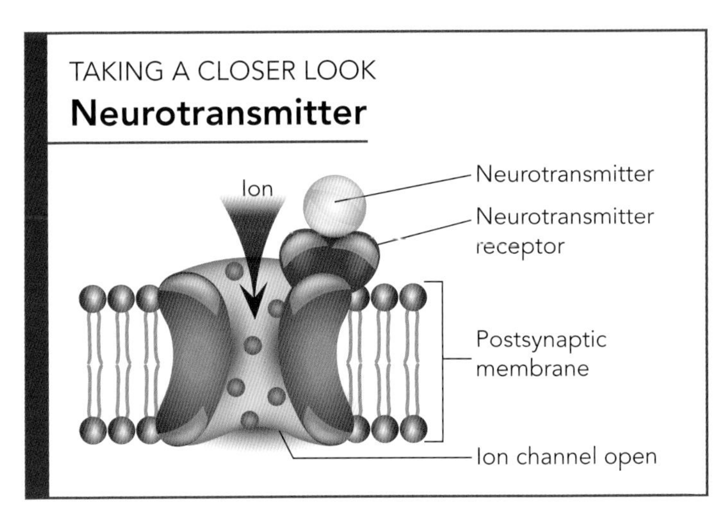

Neurotransmitters

Neurotransmitters are the molecules that carry the signals across the synaptic cleft. They then interact with the channel proteins on the postsynaptic membrane, triggering ions channels to open. But what happens after that?

The actual work of neurotransmitter lasts only for a few milliseconds (thousandths of a second). Then it disengages from its receptor protein, ending its effect on the postsynaptic cell. But the neurotransmitter is still there, correct? It needs to be removed before the next batch of neurotransmitters is released, correct? Correct...both times.

There are several processes that are in place to remove the neurotransmitter after its work is done. In some cases the neurotransmitter simply drifts away from its receptor. Special enzymes can break down specific neurotransmitters to remove them from the synaptic cleft. And some neurotransmitters are pumped back into the axon terminal of the presynaptic neuron.

There are several dozen known neurotransmitters. This list includes acetylcholine, epinephrine, dopamine, glutamate, histamine, serotonin, and gamma-aminobutyric acid (GABA). Some are associated with specific areas of the brain or spinal cord, while some are found in the peripheral nervous system. Acetylcholine, for instance, is the neurotransmitter used by nerves synapsing with skeletal muscles but is also found in the brain. And norepinephrine is used in both the brain and the sympathetic nervous system. Some neurotransmitters send excitatory messages, while others are more inhibitory, and others can either excite or inhibit depending on the type of receptor. God used a wide variety of neurotransmitters in the design of the nervous system, fine tuning its many parts to communicate efficiently with each other and with the rest of the body.

Homeostasis

As we have seen throughout our exploration of the human body, the body is constantly trying to maintain a "balance" among its many systems. This is known as homeostasis.

For the body to function properly and efficiently, its internal conditions must be kept within optimal ranges. There are numerous processes that the body must monitor and regulate from second to second, minute to minute.

The body's temperature must be kept within the correct limits, not too high, not too low. Blood pressure, heart rate, and respiration are constantly evaluated and assessed. Is the blood level of thyroid hormone appropriate? If it's too low, a person might feel sluggish. If it's too high then the body's metabolism can be accelerated. Is the level of oxygen in the blood high enough? It certainly needs to be. On the other hand, is carbon dioxide building up in the blood? If so, the respiratory system needs to be signaled to increase ventilation. And it goes on and on.

We will explore many of these processes during our studies. However, as we proceed you need to consider something. How could mechanisms so complex be an accident? The answer is that they are not. They could not possibly be the result of chemical reactions over millions of years.

The processes that aid in homeostasis bear the mark of amazing design. There is nothing random or accidental about them. They are the work of the Master Designer.

For You formed my inward parts;
You covered me in my mother's womb.
I will praise You, for I am fearfully and wonderfully made;
Marvelous are Your works,
And that my soul knows very well.
(Psalm 139:13–14)

4

THE CENTRAL NERVOUS SYSTEM

Walking, talking, sitting, standing, learning, sleeping, laughing—just imagine all the things our nervous system has to process every second to make any of these things possible. It is quite literally beyond our imagination.

It seems like the more we understand about the nervous system, the more there is to learn. How does it all work, and how does it all work together?

The vast number of neurons, the even greater number of interconnections, the complex structures, and (probably the biggest thing) all those strange names given to all this stuff...it's all just too much!

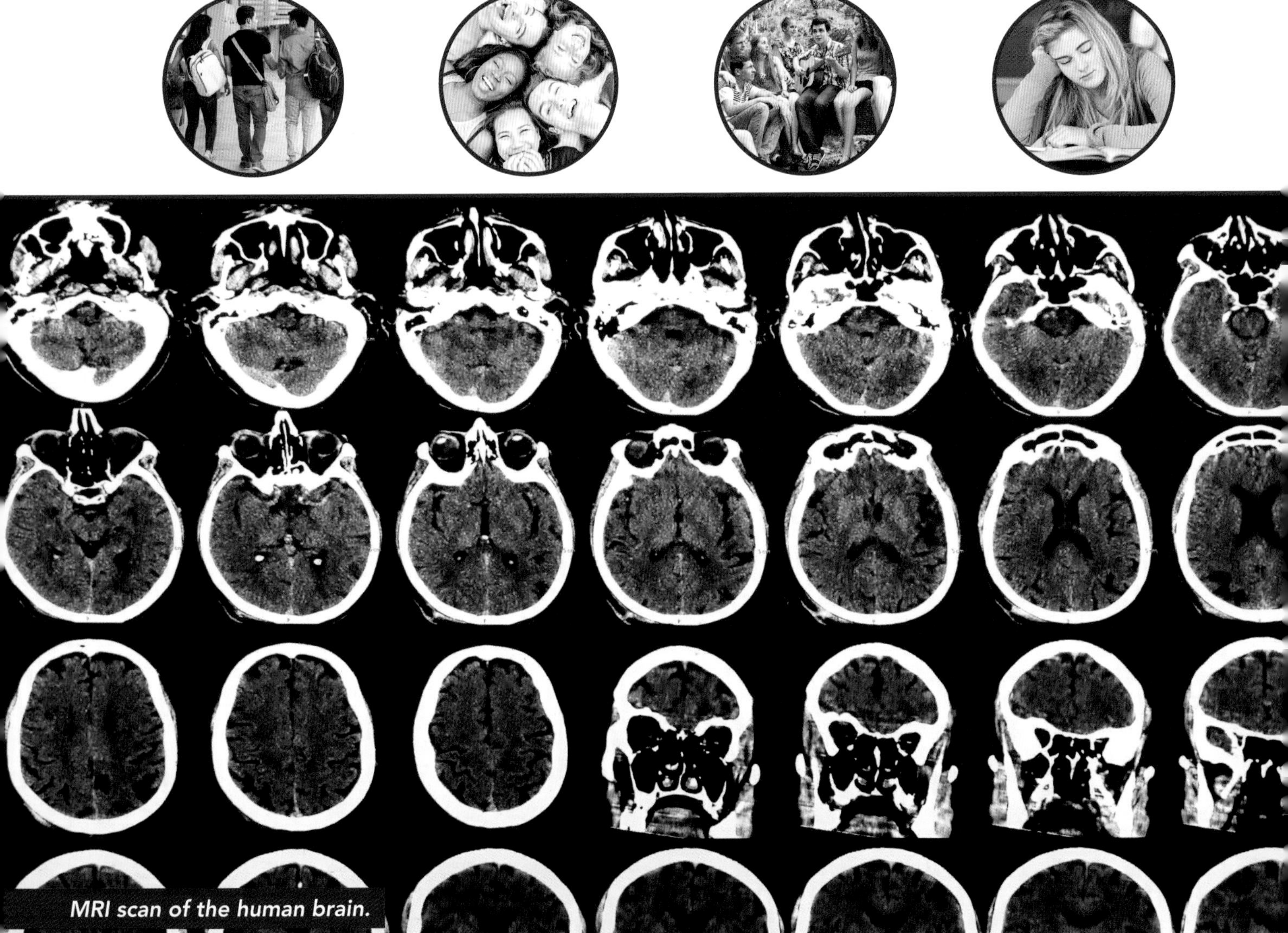

MRI scan of the human brain.

No, it really isn't. We will continue to take all this a step at a time.

Now that we've seen how nerve impulses travel, let's introduce the brain and spinal cord, the central nervous system.

The Brain

The brain is the master control center of the human body.

Standing up without falling down? The brain controls this. Can you taste that ice cream? Yep, the brain is right on the job. Make an A on that math test? Good job, brain! Feeling sleepy? Your brain is telling you something.

The most amazing, incredible, complex thing in the universe is the brain. Even with all we have learned in recent decades, it is still the least understood organ in the human body.

The brain is so unimaginably complex and so astonishing in its function, it can only be the work of the Master Designer.

The human brain is quite recognizable to most people. It is an organ that is a sort of pinkish-gray in color and has many wrinkles or folds on its surface. An average brain weighs around 3 pounds. That's approximately 2 percent of the body weight of a typical adult. Even though its size is small, the brain consumes about 20 percent of the body's total energy.

How many neurons are packed into this 3-pound energy guzzler? The most often quoted estimate of the number is 100 billion! More recent research

TAKING A CLOSER LOOK

The Brain

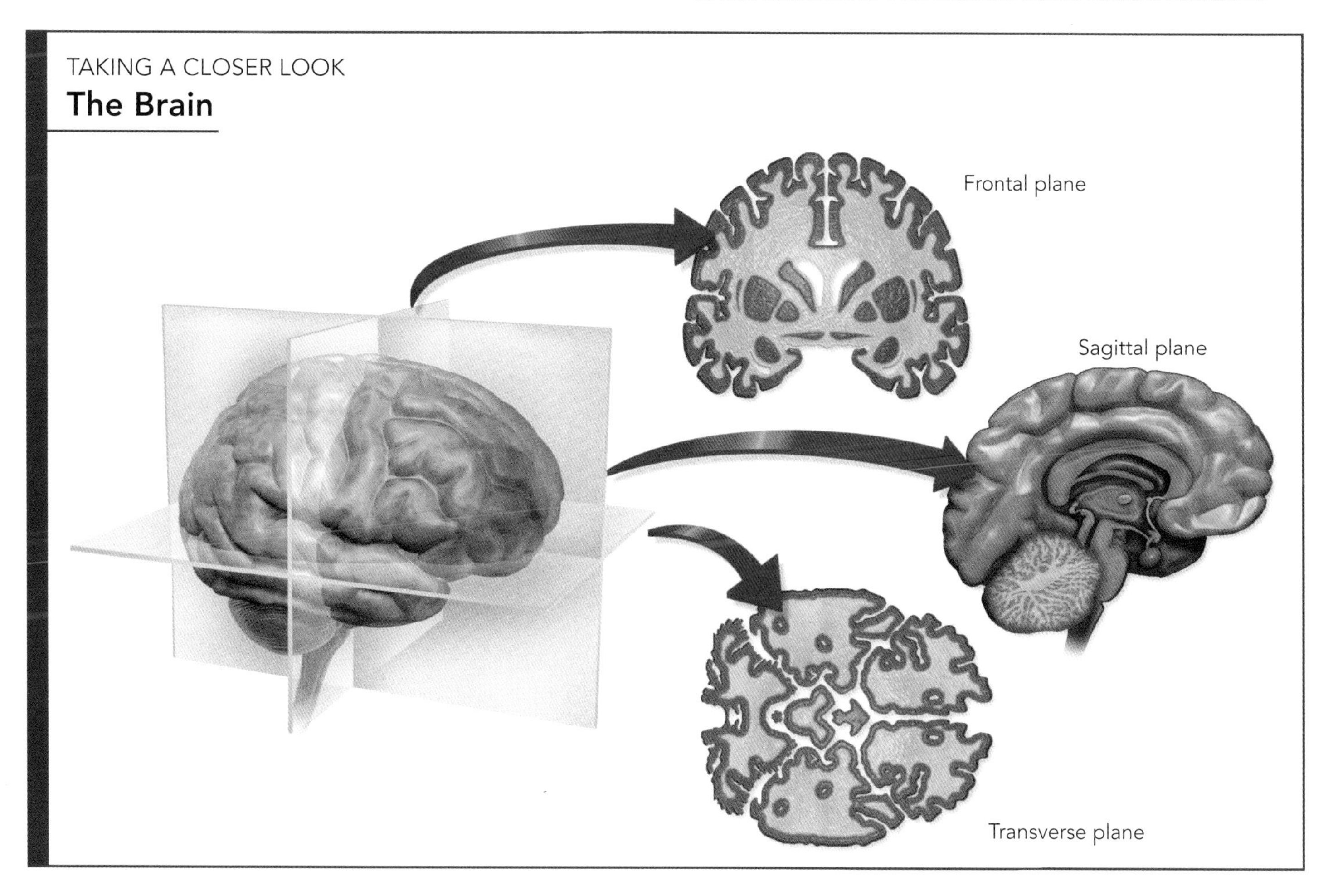

suggests that the actual number is closer to 86 billion, but that is still a lot of neurons. Each of these neurons communicates with many other neurons through synapses. The number of synapses in the brain, links between its neurons, has been estimated to be in the trillions! (Seriously, who counted them?)

Brain - Its Protection

The brain is such a vital organ, and it makes sense that it is located where it can be well protected.

The brain is located inside the cranial vault at the top of the skull. The skull consists of the cranial vault and the bones that make up your face and jaw. The cranial vault—also called the cranium—is the large open space in the skull. This bony container, shaped perfectly to house the human brain, provides significant protection from trauma.

Inside the cranium, the brain is cushioned by another layer of protection, the meninges. The meninges are really three layers of protection—three layers of connective tissue that cover the brain and spinal cord. The outermost layer is called the dura mater.

The dura (short for dura mater) is the thickest layer of the meninges and is itself composed of two layers. The outermost layer of the dura is attached to the inside of the cranium and is called the periosteal layer. (If you recall from Volume 1 of *Wonders of the Human Body* that the outer covering of bone is called the periosteum, it will help you remember that this periosteal layer is closest to the cranial bone.) The inner layer of the dura is called the meningeal layer. The two layers of the dura mater are fused together except in those areas where they separate to form

Do We Only Use 10 Percent of Our Brains?

Is it true that we only use 10 percent of our brains? You've probably been told that your whole life by people who are sure it is true because they've heard it from folks who were sure it was so.

Well, no one was trying to pull the wool over your eyes, but it is not true. Not at all.

This "10 percent myth" is one of the most common myths in our modern culture. Somehow people have come to believe that we only use a small portion of our brain every day. I have so often heard, "Wouldn't it wonderful if we could unlock that unused part of our brains?" It would be. If it were true. But it isn't.

You see, with all the modern scanning and imaging techniques available to neuroscientists, no one has been able to discover exactly where this unused part of the brain is located. Plus, scanning and mapping studies of the brain have failed to find a part of the brain that is not active, even with basic activities such as walking or watching television.

So you see, the 10 percent brain myth is 100 percent false.

TAKING A CLOSER LOOK

The Cranial Vault

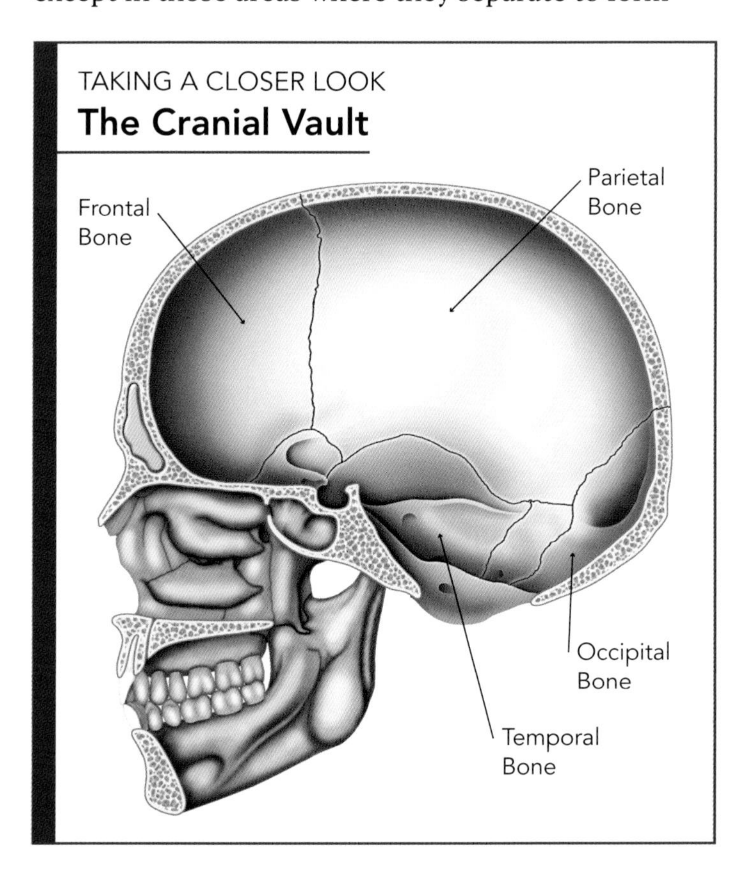

large veins called venous sinuses. These venous sinuses provide channels to drain venous blood from the brain on its way back to the heart.

The middle layer of the meninges is known as the arachnoid mater. This name might sound spidery to you, if you recall that spiders are called arachnids. Both words come from the Greek word for spider, *arachne*. From the arachnoid mater extend fine delicate fibers that resembles spider web. That's how it got its name! Like the dura mater, the arachnoid mater covers the brain and spinal cord. Neither of these layers dips down into the brain's many folds.

The third meningeal layer is the pia mater. This very thin layer of connective tissue lies next to the brain itself and is covered in very tiny blood vessels. The pia mater dips down into the folds and grooves in the brain.

The spaces between the meningeal layers also have names, naturally. The space between the dura and the arachnoid layer is called the subdural space. Sub means "under," so that's a pretty good name—subdural means "under the dura." The space below the arachnoid is called (logically enough) the subarachnoid space. And these spaces are a very important part of the cushion that protects the brain!

Cerebrospinal Fluid

In the subarachnoid space is a liquid known as cerebrospinal fluid (CSF). This fluid flows around the brain and spinal cord, cushioning both. The CSF supports the brain by helping it float in the cranial vault. It's sort of a waterbed for your brain! This fluid layer also acts as a shock absorber to keep the brain from banging around inside the skull. Simply running up and down stairs could damage the brain and the surrounding structures if there were not a way to keep the brain from moving around and bumping into the bone that houses it. Another amazing design!

In addition to the subarachnoid space, CSF fills four chambers in the brain called ventricles. The two largest are the lateral ventricles, one in each side of the brain, inside the cerebral hemispheres. (The two cerebral hemispheres, as we will see in a moment, are

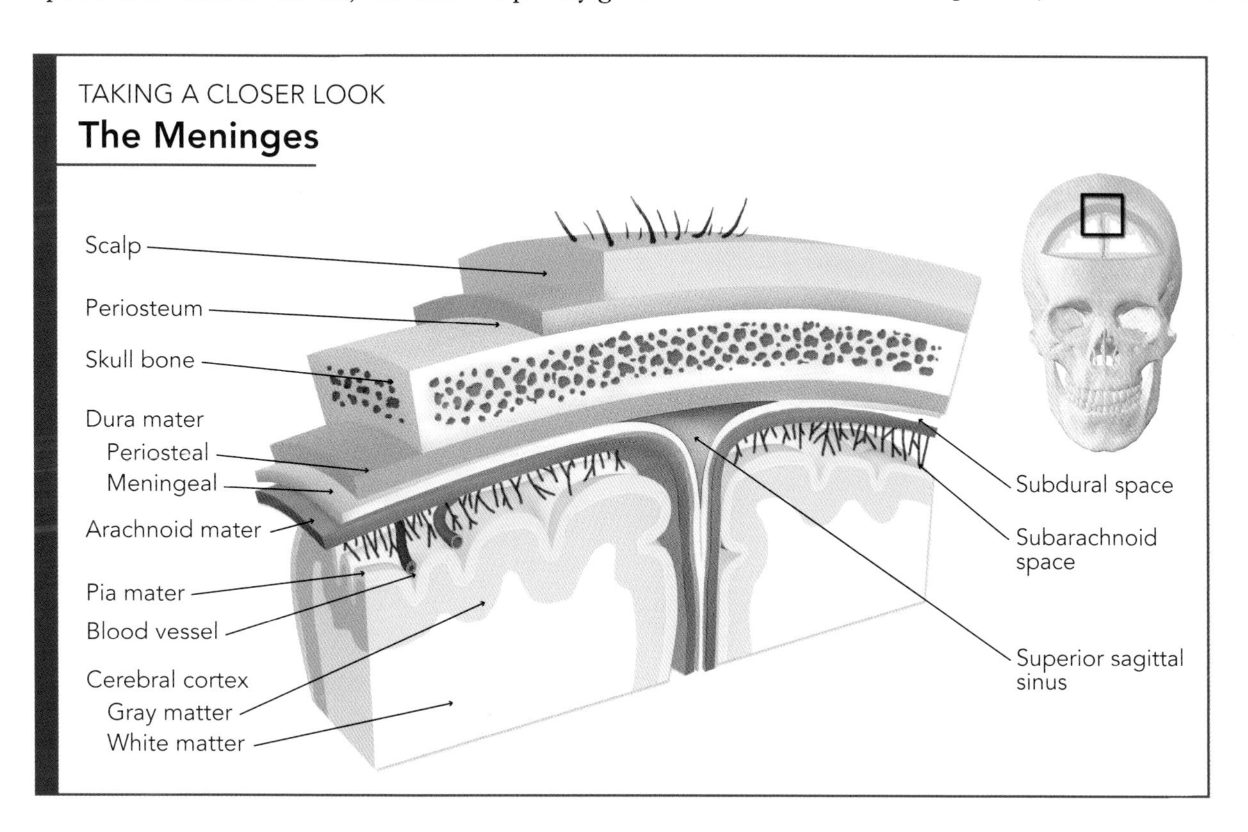

TAKING A CLOSER LOOK
Ventricles of the Brain

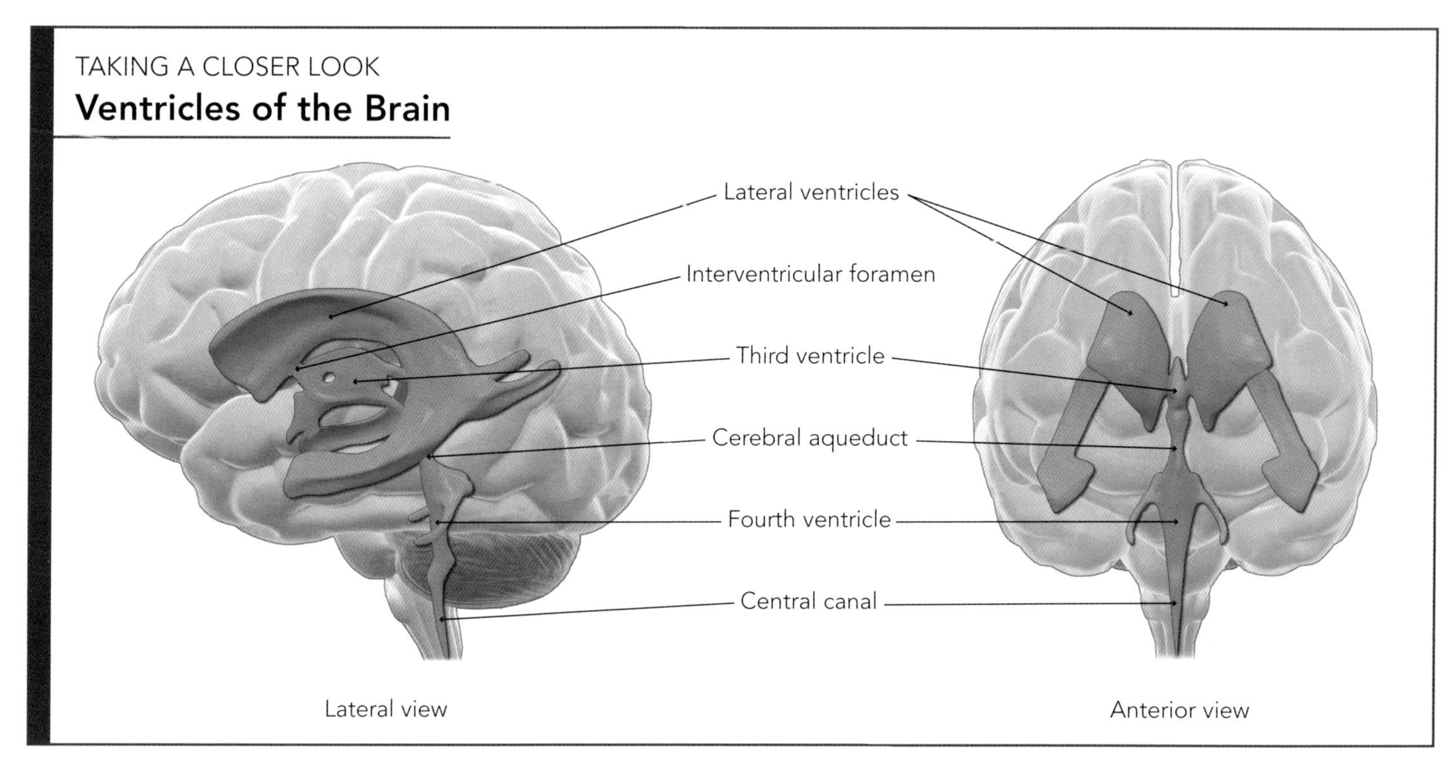

the two halves of the brain's upper part.) Between the cerebral hemispheres is the third ventricle. The lateral ventricles communicate with the third ventricle below them. Below the third ventricle is the fourth ventricle, and this ventricle connects to the central canal of the spinal cord. The interconnections between the four ventricles and the central canal allow the CSF to circulate around the brain and spinal cord.

The cerebrospinal fluid is clear and colorless. It is made from blood's clear liquid plasma, but it is completely separate from the blood. CSF is manufactured in the brain within the ventricles

Meningitis

Meningitis is an inflammation of the meninges. It is most commonly caused by an infection, either by bacteria or viruses. Because any severe inflammation so close to the brain can have dire consequences, this illness is considered a medical emergency.

Bacterial meningitis can be fatal if untreated, so prompt intervention is crucial. On the other hand, viral meningitis has no specific treatment and generally resolves on its own. It is rarely fatal.

The most common symptoms of meningitis are severe headache, fever, and a stiff neck.

Meningitis is diagnosed by inserting a long needle into the spinal canal in the lumbar region of the spine. (This is below the spinal cord itself, so the spinal cord is in no danger.) A sample of cerebrospinal fluid is withdrawn and sent for immediate analysis. The cells present in the CSF are counted and examined. If bacterial meningitis is suspected, antibiotics are administered to the patient intravenously immediately after the CSF sample is taken, even before the test results are available, because timely treatment is so important. If the CSF sample does not show evidence of bacterial meningitis, no further antibiotics are given.

TAKING A CLOSER LOOK

Choroid Plexus Section

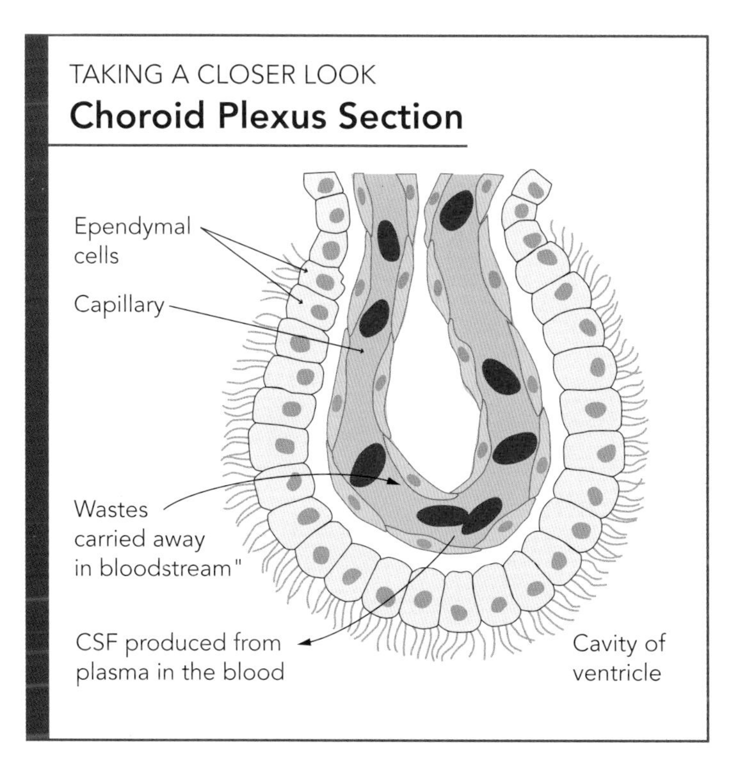

and subarachnoid space. The majority of the CSF is produced by the frond-like choroid plexus found in each ventricle. In these choroid plexuses, ependymal cells (remember they are a type of neuroglia cell) filter the plasma from the blood to produce the CSF.

Parts of the Brain

The human brain is made of four major parts. The largest part is called the cerebrum, which is made of two halves, the cerebral hemispheres. Next is the diencephalon. The diencephalon is made up of the structures near the third ventricle—the thalamus, hypothalamus, and the epithalamus. (We'll get to what those parts do later.) At the rear of the brain, tucked under the back of the cerebrum, is the cerebellum. And on the underside of the brain is the brain stem. The brain stem ultimately merges with the spinal cord.

Let's explore the structure and then the functions of each part in turn.

Cerebrum — Gross Anatomy, the Parts

The most recognizable portion of the brain is the cerebrum. It is somewhat wrinkled in appearance, covered with many ridges and folds. The ridges are called gyri (singular is gyrus), and the folds are called

TAKING A CLOSER LOOK

Brain Structures

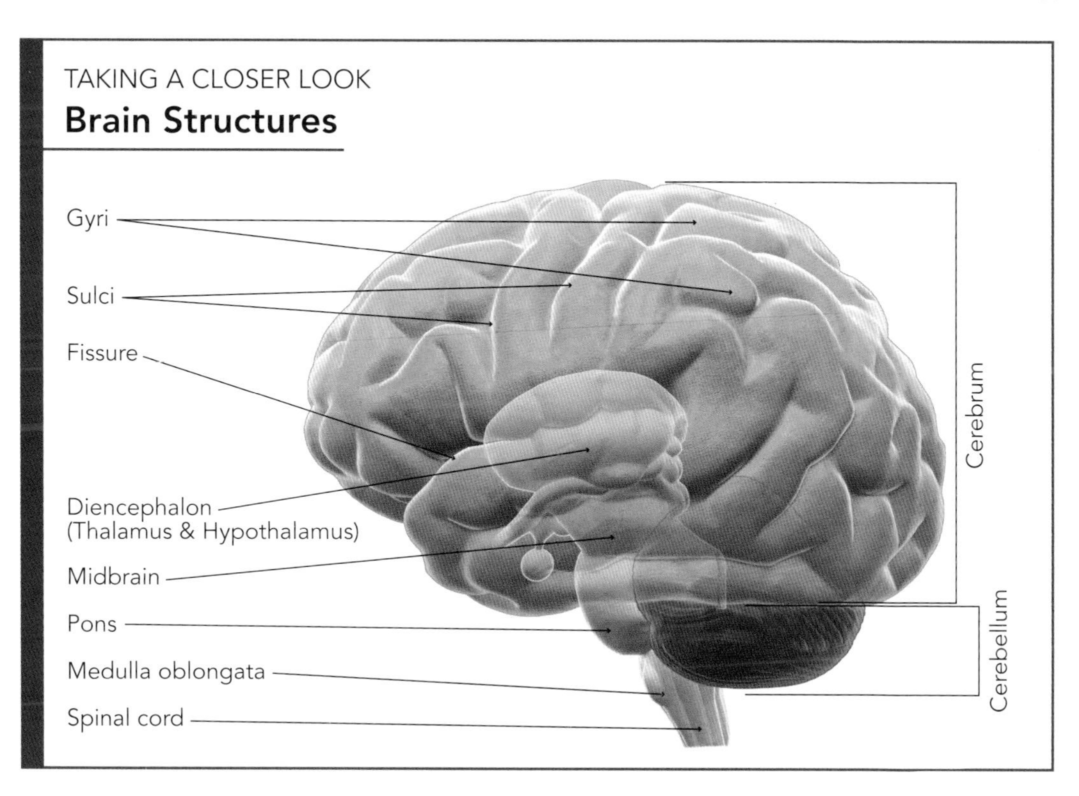

sulci (singular is sulcus). In addition, there are a few deeper folds called fissures. The fissures separate major portions of the cerebrum.

If you examine the illustration provided, you can see that there are two halves to the cerebrum. These are called the cerebral hemispheres. These two hemispheres are separated by a large fissure known as the longitudinal fissure.

Several smaller fissures divide each hemisphere into smaller portions, called lobes. In the front of each hemisphere is the frontal lobe. Moving posteriorly (toward the back), you find the parietal lobe. The frontal and parietal lobes are separated by the central sulcus. Then, continuing posteriorly, separated from the parietal lobe by the parieto-occipital sulcus, is the occipital lobe. Below the parietal lobe is the temporal lobe. This lobe is bordered by the lateral sulcus. Make sure you can locate each of these lobes and the large sulci and fissures that demarcate them in the illustrations.

TAKING A CLOSER LOOK

Lobes of the Brain

Anterior
Posterior
Anterior
Frontal lobe
Parietal lobe
Occipital lobe
Temporal lobe
Posterior
Left
Right

Cerebrum — Gross Anatomy — the White and the Gray

Let's now examine a cross section of the cerebrum. The illustration below is a "frontal section," which is a cross section made by separating the front and back parts of a structure. Notice the cerebrum's gray outer layer. Have you ever heard of "gray matter"? Gray matter is made up of the cell bodies of neurons and neuroglia. And this gray matter is the cerebral cortex.

Deeper inside notice the cerebral white matter. White matter is made up of both myelinated and nonmyelinated axons.

We can get a glimpse deeper into the brain from a "sagittal section." The sagittal section on the next page is a cross-sectional view we get when we separate the right and left parts of a structure. The most important structure to note here is called the

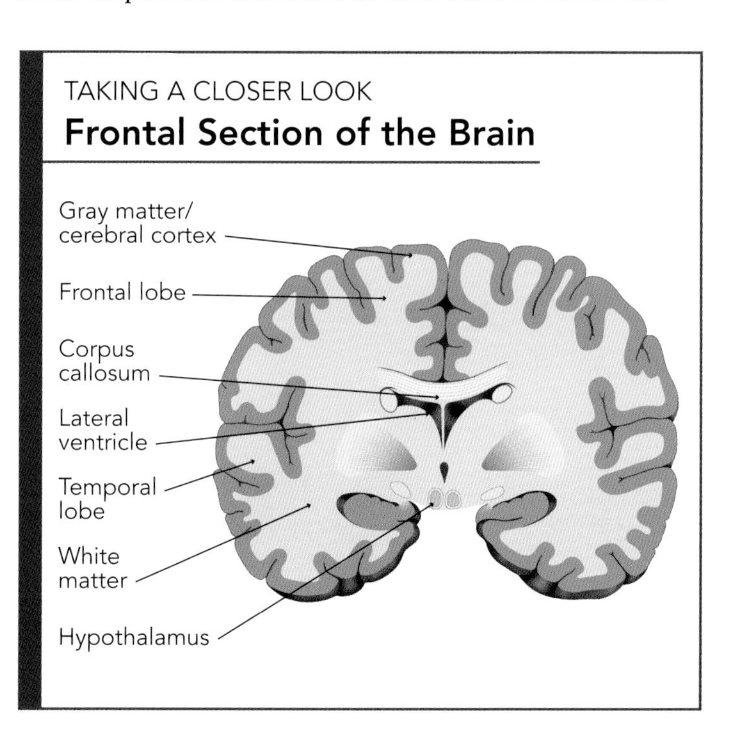

corpus callosum. The corpus callosum is a large band of white matter that connects the two cerebral hemispheres. Thanks to the axons that cross from one side of the cerebrum to the other through the corpus callosum, the two cerebral hemispheres communicate and work together.

That's enough for now about the gross anatomy of the cerebrum. We will mention a few other very important regions later, as we learn just what the cerebrum does.

Cerebrum — Motor Functions

The cerebrum controls so many things in the body it is difficult to know where to begin. It controls voluntary movement, the movement you can consciously control. The cerebrum enables you to perceive the world by making you aware of the sensations you receive every second. The centers for speech and language are in the cerebrum. Signal processing in the cerebrum allows us to learn and understand. We can remember the things we learn and experience because the cerebrum helps store memories. The list could go on, but you get the idea.

TAKING A CLOSER LOOK

Sagittal Section of the Brain

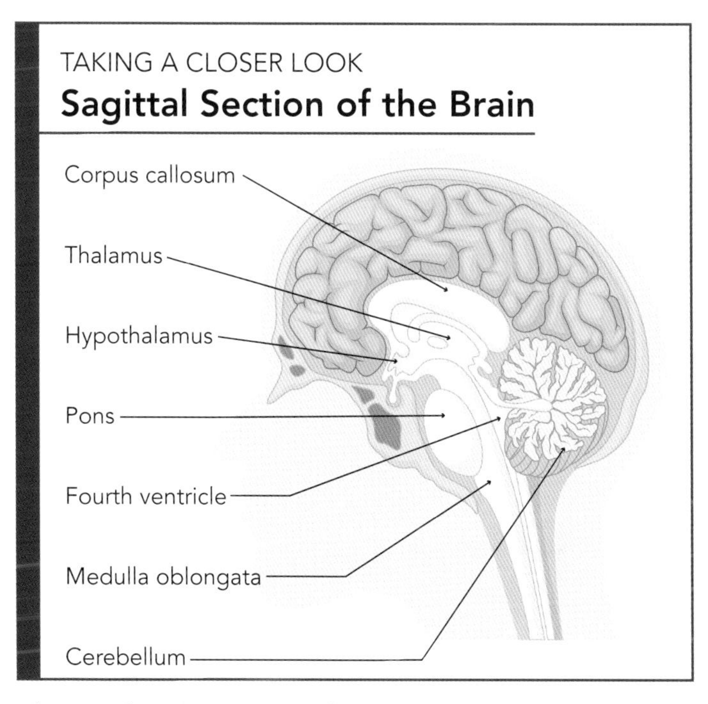

The cerebrum consists of the two cerebral hemispheres. Each hemisphere is equally important, but they do not necessarily have all the same jobs. Each of the cerebral hemispheres controls one half of the body. Interestingly enough, they control the opposite side of the body. That is, each cerebral hemisphere controls the motor functions on the opposite side of the body. Each cerebral hemisphere

TAKING A CLOSER LOOK

Cerebrum

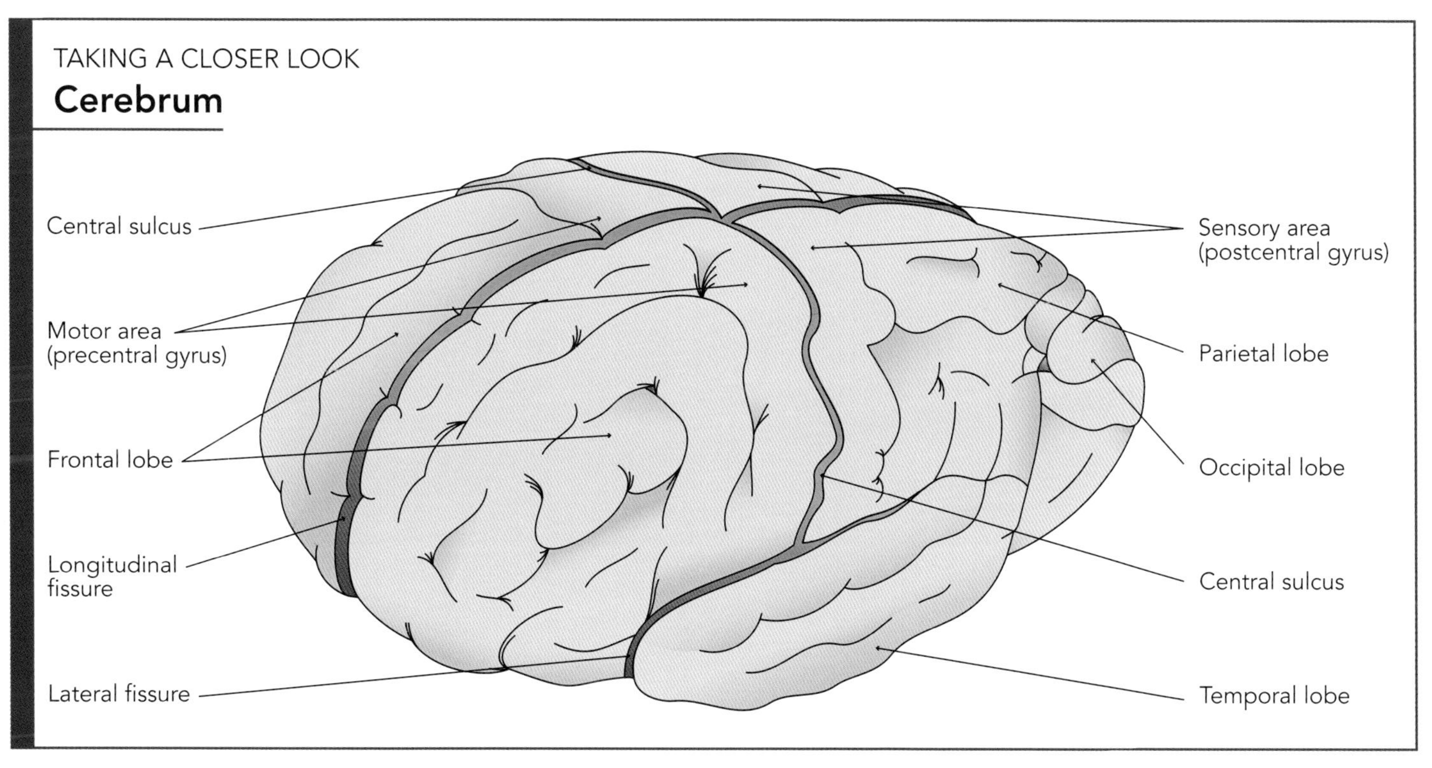

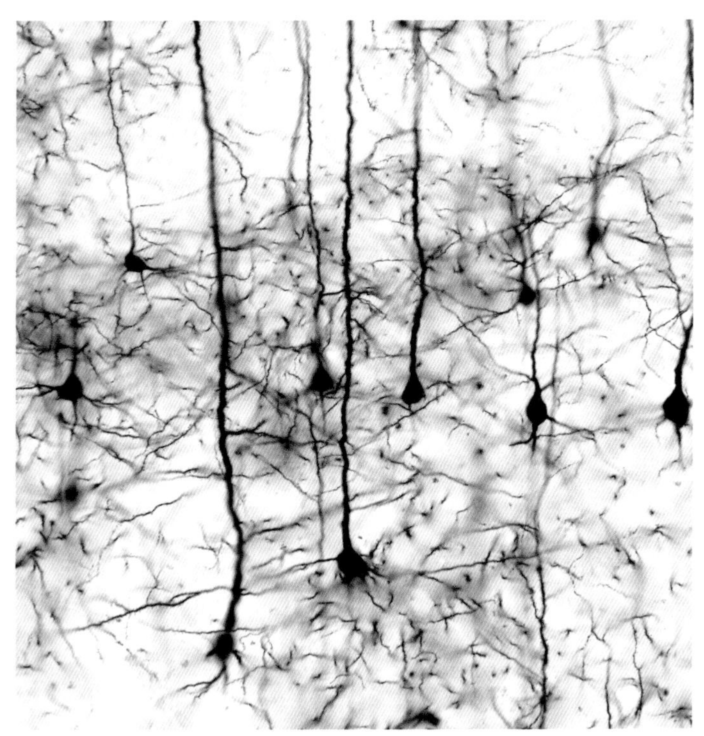

Pyramidal neurons of the cerebral cortex.

also receives the sensory input from the opposite side of the body. Thus the right hemisphere controls and monitors the left side of the body, and the left hemisphere controls and monitors the right side of the body!

There are, however, some functions of the brain that are lateralized. That is, the responsibility for certain functions rests with one hemisphere or the other. We will see examples of these as we go on.

Finally, and perhaps most importantly, you need to understand that nothing that happens in the cerebrum, happens in isolation. The brain's neural pathways are very, very highly integrated—interconnected and coordinated. The signals into and out of the brain are constantly enhanced, inhibited, and modulated (changed) by input from all across the cerebrum. Thus, even while we describe the brain's structures and functions as simply as possible, we understand that most things about the brain are really more complicated than they appear on the surface.

Motor function is controlled by a region of the cerebrum called the primary motor cortex. The primary motor cortex is located on the precentral gyrus. This is the ridge just in front of the central sulcus. Pre- means "in front of," and gyrus means "ridge," so the name precentral gyrus makes sense: "the ridge in front of the central sulcus."

The primary motor cortex is made of a special type of neuron called pyramidal cells (or pyramidal neurons). Pyramidal neurons have very long axons that extend all the way into the spinal cord. Bundles of these axons are called pyramidal tracts. You probably recall that in the PNS bundles of axons were called nerves. Well we aren't in the peripheral nervous system now. In the central nervous system, bundles of axons are called tracts. The pyramidal tracts carry the message "move!" from the primary motor cortex to the spinal cord. (Watch for the pyramidal tracts to bulge into view again when we discuss the brainstem.) From the spinal cord, the "move" message is sent to the designated muscles to get the job done.

So how does the brain know to move my leg when I walk? How does it know how to send messages to the correct fingers when I type?

As it happens the primary motor cortex has specific areas that correspond to specific parts of the body. Over many years these areas have been mapped and studied by neuroscientists. You can see the various parts of the body mapped out in the illustration here. This type of visual representation of the motor regions is called a motor homunculus (meaning "little man"). The motor cortex maps in a similar way in each hemisphere. (But remember, left brain controls right body, and right brain controls left body. Right? No, left...seriously?)

Examining the homunculus, you see that it pictures the body upside down. The feet and legs are at the top, and the face is at the bottom. This doesn't mean the brain considers the feet more important than the face. Far from it. As you look more closely, you will find that some body regions take up more of the motor area than others. In fact, the hand and face take up the majority of the motor cortex. Completely logical, right? The hands require very precise motor control, as do the muscles of the face and jaw. Writing, typing, playing an instrument, smiling, talking, chewing...all require lots of muscles acting in just the right way. We aren't surprised then that God designed the brain with a lot of territory devoted to muscles in the hands and face.

But, as you might have guessed, muscular control is more complex than the homunculus illustration suggests. This map of the "little man" gives you a general idea of where the major control areas are, but voluntary muscle movements are fine-tuned by a vast number of neural inputs from areas throughout the cerebral cortex.

When studying the control of motor function, there is another very important region to examine. This is the area just anterior to (in front of) the primary motor cortex. Remember pre- means "in front of." Therefore this important part of cerebral cortex—"in front of the motor cortex"—is called the premotor cortex. Here all the planning happens before any movement begins! You see, something like reaching for a glass or picking up a pencil is not as easy as it might seem. The coordination of many muscles is necessary to perform even relatively simple tasks. The preparation for these movements occurs in the premotor cortex, even though you don't often realize you are making plans for how to move. Which muscles must move, and in what order they move are all laid out in the premotor cortex. The premotor

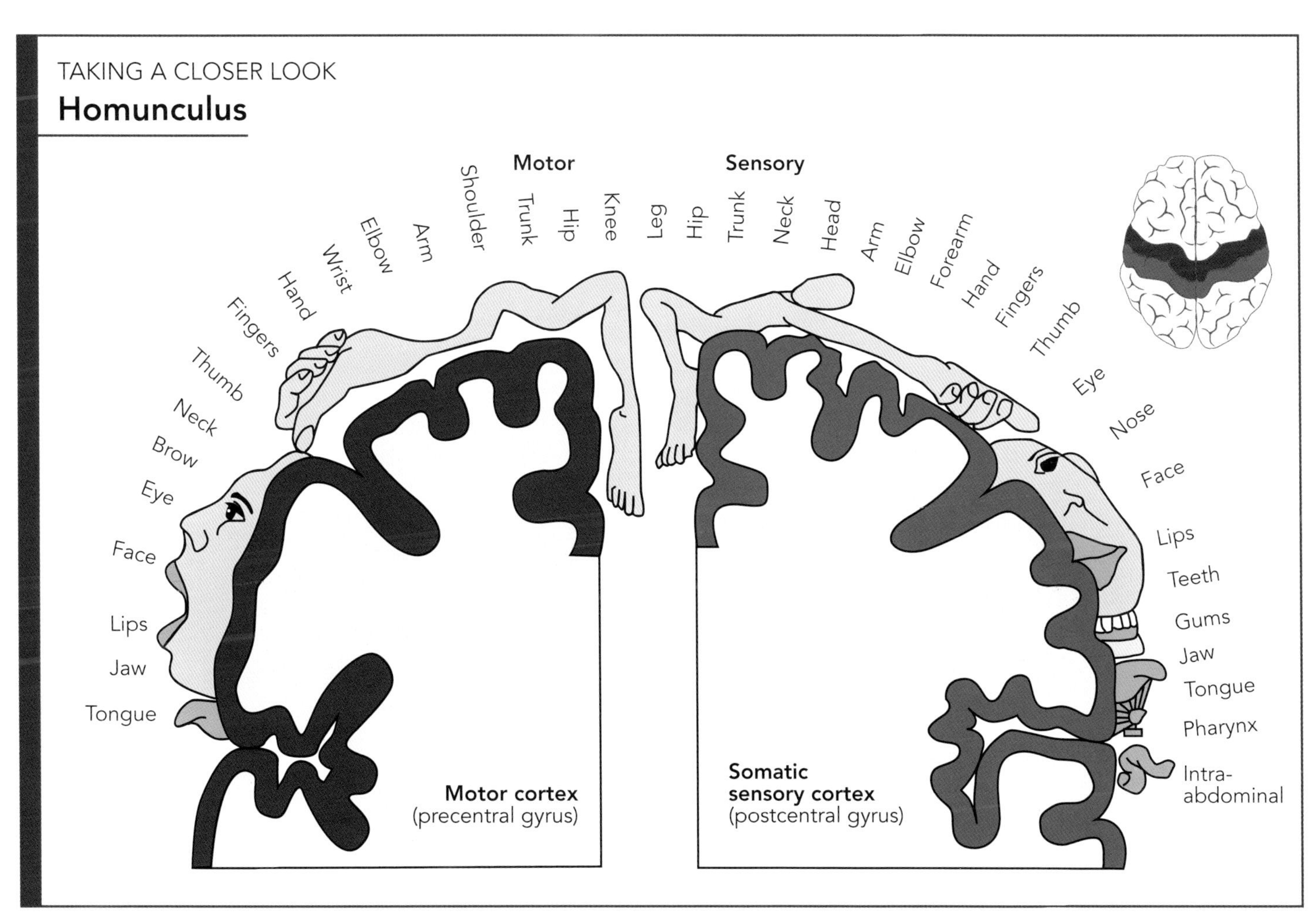

cortex then sends appropriate signals to the primary motor cortex to get voluntary muscle movement (or movements) underway.

In addition to the areas we have covered thus far, there is one more very interesting region to examine. It is known as Broca's area. In most people, Broca's area is located in the left cerebral hemisphere, though in some folks Broca's area is on the right. Broca's area is thus lateralized. It is found in only one of the hemispheres.

Broca's area controls muscles involved with speech. Muscles in the mouth and larynx are coordinated by Broca's area, enabling us to speak. Neurons here interact with both the premotor cortex and the primary motor cortex to manage this complex process of communicating using spoken language.

The Cerebrum — Sensory Functions

The cerebrum is also involved with processing sensory information from throughout the body. Sensory information is processed in the primary somatosensory cortex. This enormous name makes sense, if we break it down. Primary means is the area first in importance for this function. Somato- means "body." And sensory, well that means having to do with input to the brain, such as information from your sensory organs.

The primary somatosensory cortex is located in the parietal lobe of the cerebrum on the postcentral gyrus. Does that name tell you where it is located? If you said, "behind the central gyrus," (because post- means "behind"), then you are catching on to the pattern behind these big names.

The cerebrum's primary somatosensory cortex receives sensory inputs from all over the body. Touching a piano key, sitting in a chair, drinking warm cocoa, stubbing your toe...all these actions produce sensory inputs that are ultimately processed in the primary somatosensory cortex.

TAKING A CLOSER LOOK

Broca's Area

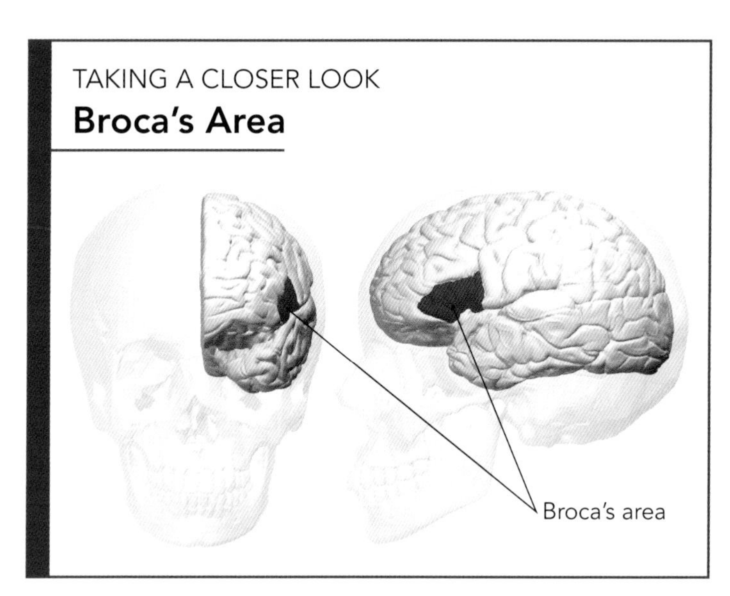

AND...just as the primary motor cortex has specific areas that send signals out to specific areas of the body, the primary somatosensory cortex has specific areas that receive input from specific areas of the body. Just as with the primary motor cortex, neuroscientists have mapped the primary somatosensory cortex. This mapping can be represented by a somatosensory homunculus, another "little man" map. Once again, the senses on the hands and lips map onto a greater territory than touch sensations from other parts of the body. The primary somatosensory cortex maps in a similar way in each hemisphere.

Expressive Aphasia

As we have seen, Broca's area controls the muscle activity in involved with speech. When people suffer damage to Broca's area, perhaps after a stroke, they often lose the ability to speak. In this very unfortunate situation, people can comprehend and process everything that is said to them. And they know what they want to say, but they are not able to make the appropriate muscle movements to form words. This inability to speak, to "express themselves," is called expressive aphasia.

Remember that motor functions start by being planned in a special area (the premotor cortex) before instructions to move something are issued by the primary motor cortex. We see the same pattern on the sensory side of things.

Just behind the primary somatosensory cortex is the somatosensory association cortex. This area is a processing area for sensory inputs. Here many sensory inputs are received and integrated. This signal processing allows the cerebrum to take many individual inputs and understand their greater meaning. Because the somatosensory association cortex combines, evaluates, and interprets a variety of signals and signal types, our brain can, for instance, distinguish whether a round object in your hand is a marble or a grape, even without looking. Sensory integration is incredibly complex, but we need to praise God that our brains have this ability. We could not make sense of our world without it.

Cerebrum — Association Areas

Other important areas of the cerebrum receive and process information from many sources. These are known as association areas. There are many of these, but here we only consider a few:

The frontal association area is located in front part of the frontal lobe. This extraordinarily complex area controls many of our intellectual functions. It is involved with learning, reasoning, planning, and abstract reasoning (you know, important questions like, "Why did the chicken cross the road?"). Many aspects of our personality are determined by the frontal association area.

The visual association area is in the occipital lobe. This area processes visual information to allow us to understand what we are looking at. Are we staring at a bee or a butterfly? When you are looking for a snack, be thankful that your visual association area helps you distinguish between an apple and an onion.

Wernicke's area is found in the temporal and parietal lobes of the left cerebral hemisphere. It is in Wernicke's area that the language we hear is processed, allowing us to understand speech. Wernicke's area, like Broca's area, is a lateralized part of the brain associated with language.

The auditory association area is in the temporal lobe. This area helps us distinguish between types of sounds. Are we hearing

TAKING A CLOSER LOOK

Association Areas

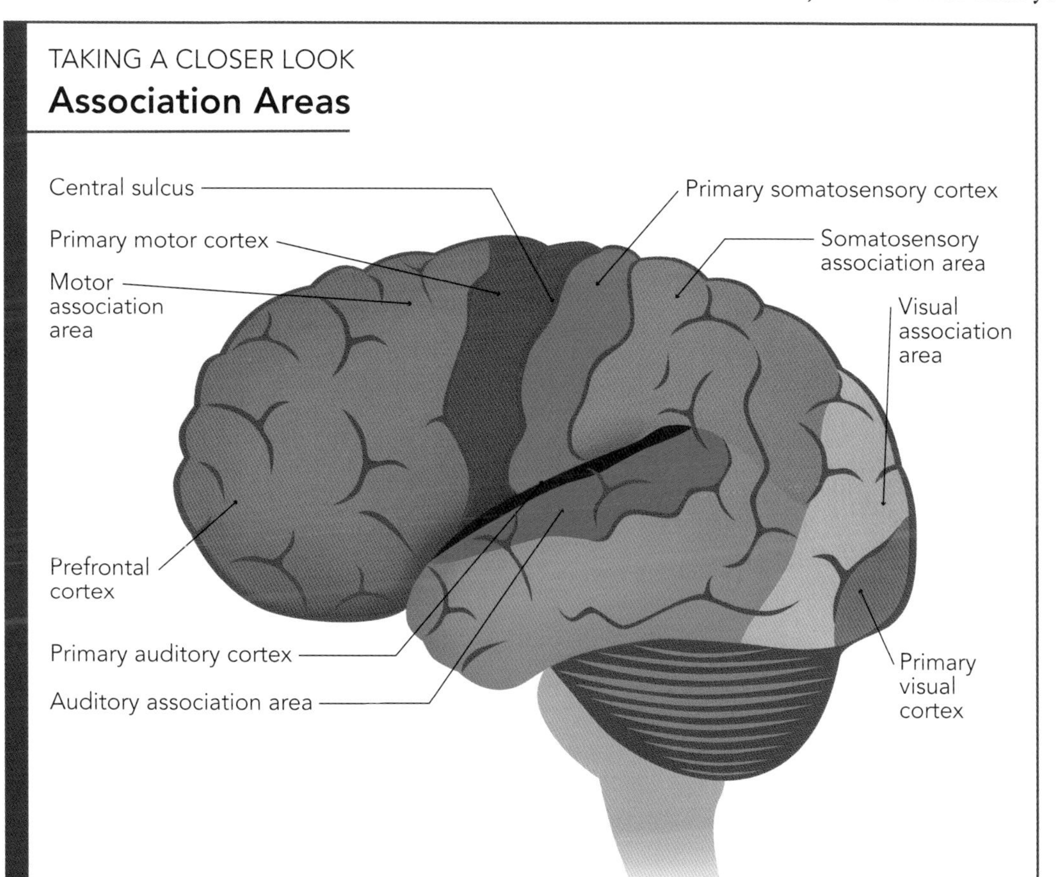

Right-Brained? Left-Brained? Scatter-Brained?

In recent years much has been written about being "right-brained" or "left-brained." This idea is the result of many popular misconceptions about how the brain works.

It has been said that the right hemisphere of the brain is devoted more to emotions, music, pattern recognition, and artistic endeavors. At the same time, it said that the left hemisphere is the seat of logic, reasoning, and planning. To a certain degree, these things are true, as certain functions in the brain are lateralized.

This discovery has caused people to claim that artistic and thoughtful people are "right-brained." At the same time, people who are analytical and logical are called "left-brained." In the final analysis, this is a gross oversimplification. No one is truly that "dominant" on either side of the brain. In fact, we all do much better when we use "all" of our brain at the same time (Some people don't. . . . We call them "scatter-brained.")

To further dispel this myth, let me use myself as an example.

Over the years, I have felt myself to be a reasonable (though not gifted by any means) guitar player. Thus, I am a musician. Also, my primary hobby for over 30 years has been photography. Therefore I am an artist. That would mean I am "right-brained," right?

On the other hand, for over 20 years, I worked in an intensive care unit taking care of very sick people. In that setting logic, reason, and critical thinking are first and foremost. Since I took good care of my patients, I must be "left-brained," right?

Well both can't be true. Apparently this left-and right-brained stuff isn't so straightforward as popular wisdom would suggest.

Yeah, I know what you are thinking. Scatter-brained, right?

music or laughter? Are we hearing thunder or a firecracker? Like other association areas, this one helps us interpret and understand our world.

Which Is the Important Side?

So which side of the brain is the most important? That's easy. The answer is BOTH!

There really is no dominant side of the brain. One cerebral hemisphere is not more important than the other. Remember that the left cerebral hemisphere gets sensory signals from the right side of the body. It also controls the voluntary motor function of the right side of the body. The right cerebral hemisphere gets sensory input from the left side of the body and controls voluntary muscle movement on the left side. So asking which side if the brain is more important is like asking which side of the body is more important.

It is true that certain functions are performed by either one hemisphere or the other, Broca's area, for example. However, just because some specific things are lateralized in the brain does not mean that side of the brain is more important. The white matter bands in the corpus callosum connect the right and left hemispheres to allow communication between the hemispheres. This communication is very, very important. With essentially every process, the two hemispheres are in constant communication, assisting and assessing each other.

Speaking and understanding language are lateralized functions of the brain. And the precise hand control seen in a person's dominant hand is achieved, you remember, from control from the opposite side of the brain. Does this mean that one side of the brain is more important for those uniquely human characteristics and abilities? Let's explore that issue.

In the vast majority of people, the left hemisphere is predominant when it comes to speech and language processing. Speaking of course requires precise control of all the muscles used to produce sound. Since dominant handedness also indicates superior motor control of one hand, some claim there is a relationship between language and handedness. They claim that being right handed indicates a dominant left hemisphere. If that were always true, then logically in a left-handed person, the right hemisphere would be dominant. In fact, your author, who is left handed, often makes the statement that he is "left handed, but in my right mind." (Everyone agrees but my wife, sad to say.)

In the final analysis, it is just is not that simple. One study has shown that 95 percent of right-handed people, whose dominant hand is controlled by their left hemisphere, have language dominance in the left hemisphere. The same study indicates that only 18 percent of left-handed people, whose dominant hand is controlled by their right hemisphere, have language dominance in the right hemisphere. So handedness, as with everything else about the brain's function, does not lend itself to simple explanations. And claims that one side of the brain is more important than the other collapse when the facts are examined.

Diencephalon

The diencephalon is the portion of the brain between the cerebrum and the brainstem. It surrounds one of the fluid filled chambers we mentioned earlier, the third ventricle. The two main parts of the diencephalon are the thalamus and the hypothalamus.

The thalamus makes up 80 percent of the diencephalon. The thalamus consists of two egg-shaped masses (called thalamic nuclei) connected by stalk. The thalamus plays several major roles in the brain. First of all, the thalamus relays sensory input from the spinal cord to the primary somatosensory cortex. Next, it facilitates the transmission of signals from the cerebellum to the motor cortex. Further, it plays a role in regulating consciousness. It many ways it acts as a gatekeeper for the sensory and motor cortices.

The hypothalamus is located below the thalamus. The hypothalamus controls many body functions that seem fairly automatic to us. It is one of the main regulators of homeostasis. If you remember from earlier volumes in the *Wonders of the Human Body*, homeostasis is the body's tendency to maintain internal balance. Diverse mechanisms work together to regulate conditions that must remain fairly steady in a healthy living body. These include temperature, blood pressure, balanced levels of

TAKING A CLOSER LOOK

The Diencephalon and Brain Stem

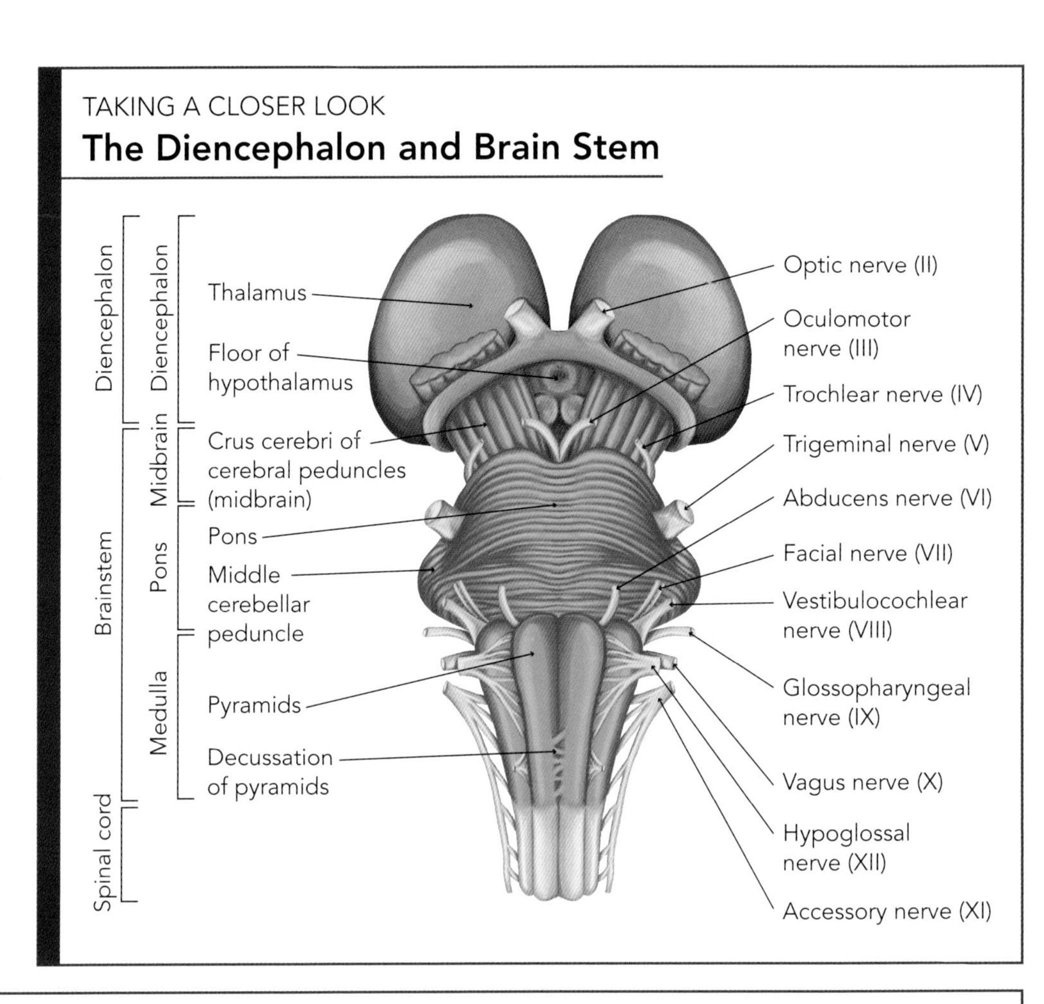

TAKING A CLOSER LOOK

Hypothalamus and Its Function

Hypothalamus
Pituitary gland
Paraventricular nucleus
Ventromedial nucleus
Anterior nucleus
Mammillary body
Preoptic nucleus (medial)
Supraoptic nucleus
Suprachiasmatic nucleus
Optic chiasm

HYPOTHALAMIC NUCLEI	FUNCTION
Paraventricular nucleus	Thyrotropin-releasing hormone release, corticotropin-releasing hormone release, oxytocin release, somatostatin release
Suprachiasmatic nucleus	Circadian rhythms
Ventromedial nucleus	Neuroendocrine control
Anterior hypothalamic nucleus	Thermoregulation, sweating
Mammillary body	Memory
Supraoptic nucleus	Vasopressin release, oxytocin release
Posterior nucleus	Increase blood pressure, pupillary dilation, shivering
Tuberomammillary nucleus	Attention, wakefulness, memory, sleep
Arcuate nucleus	Growth hormone-releasing hormone (GHRH), feeding

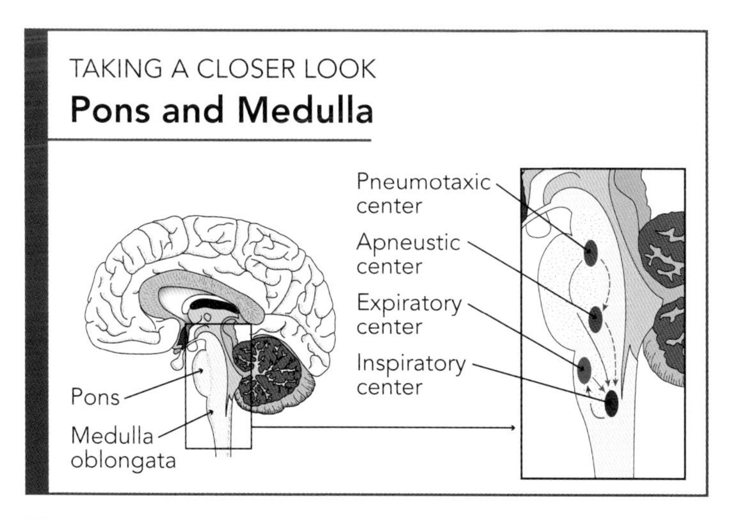

TAKING A CLOSER LOOK

Decussation

Motor area of cortex

Pons

Decussation of pyramids (crossing over)

Lateral corticospinal tract (LCST)

Anterior corticospinal tract (ACST)

Motor neuron

many hormones, and the concentrations of various substances in the blood.

The hypothalamus receives sensory input from many places throughout the body. It regulates your body's temperature. It tells you when you are thirsty. It helps sense when you are hungry and when you are full. Your attentiveness is affected by the hypothalamus, as are your sleep cycles. Heart rate, blood pressure, and even the movement of food through your digestive tract are subject to control by the hypothalamus.

Brain Stem

Proceeding downward from the diencephalon, we come to the brain stem. The brain stem consists of three parts: the midbrain, the pons, and the medulla oblongata. The brain stem provides a path for fibers extending into the spinal cord. Also, many vital body functions are controlled or regulated in the brain stem.

The midbrain is located between the diencephalon and the pons. It is roughly one inch long. It consists of two bulges in the front called cerebral peduncles. The peduncles are bundles of axons. In the rear are four nuclei known as colliculi. (In the CNS, a collection of neuron cell bodies is called a nucleus; the plural is nuclei) The colliculi are involved in both hearing and vision.

The pons (meaning "bridge") is positioned between the midbrain and the medulla oblongata. It name—bridge—reveals its function. The pons serves as a bridge linking various parts of the brain together. Through the pons run fibers from the spinal cord up into the brain and also from the cerebellum to the motor cortex. In addition there are areas in the pons that assist in the control of breathing and balance.

The medulla oblongata is below the pons, where it connects the brain to the spinal cord. Through the medulla ascend all the sensory tracts going to the brain. These are called ascending tracts. (Remember, bundles of axons are not called nerves in the CNS; they are called tracts.) Also descending through the medulla are all the motor tracts going down into the spinal cord. These are called descending tracts. The descending tracts form bulges on the anterior part of the medulla. Remember the pyramidal cells in the primary motor cortex? These bulges in the medulla contain the pyramidal tracts, the axons from those pyramidal cells. The bulges themselves are called pyramids.

Remember that each half of the cerebrum controls the muscles on the opposite side of the body. Therefore, just before reaching the spinal cord, most fibers in the pyramidal tracts decussate (cross over) to the opposite side. After this they descend into the spinal cord. This is how the right side of the brain controls the left side of the body, and the left

Headache

Headache is one of the most common complaints in modern society. Some studies have estimated that nearly 50 percent of adults have at least one headache a year. There are many different types of headaches, but here we will concern ourselves with just two: tension headaches and migraine headaches.

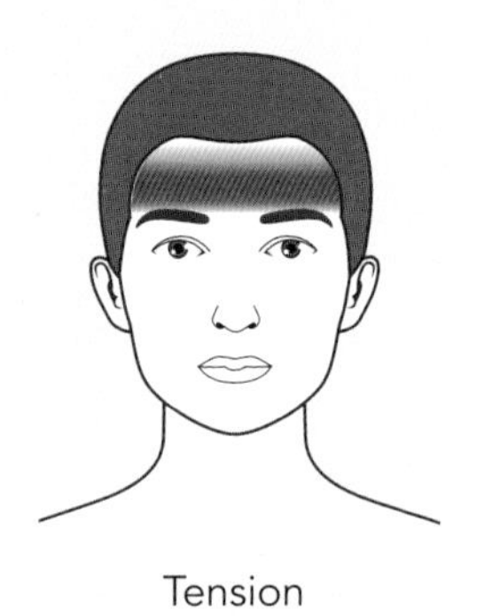

Tension

Tension headaches are the most common type of headaches. The pain of a tension headache can be mild to very severe. The discomfort can involve the sides and back of the head, the neck, and eyes. Very often the pain occurs on both sides of the head at the same time.

The most commonly cited cause of tension headaches is stress. Lack of sleep, eyestrain, and hunger can also play a role. Muscle tension in the head or in the neck often contribute.

Treatment of tension headaches is usually straightforward. In most cases a mild analgesic (pain medication) is all that is required. In the case of patients with more frequent or more severe tension headaches, a combination of medications may be required.

Migraine headaches are generally (but not always) more severe than tension headaches. Migraine pain usually affects one side of the head. It is often described as a throbbing, pounding pain. Migraines are frequently associated with nausea, vomiting, and extreme sensitivity to light and sound. It is not uncommon for someone with a migraine to retreat to a dark, quiet room to try and ease the pain. Some migraines are preceded by a sensation of seeing flickering lights that partly obscure normal vision. These are called scintillating scotoma.

The cause of migraines in unknown, although there are several schools of thought. Some researchers feel that migraines are linked to dilation of blood vessels in the brain. Others feel that dilation of arteries outside the cranium are the root cause. Some scientists feel that over-excitable neurons in the brain play a role in migraine headaches. Most likely all of these play a role. At present research is ongoing.

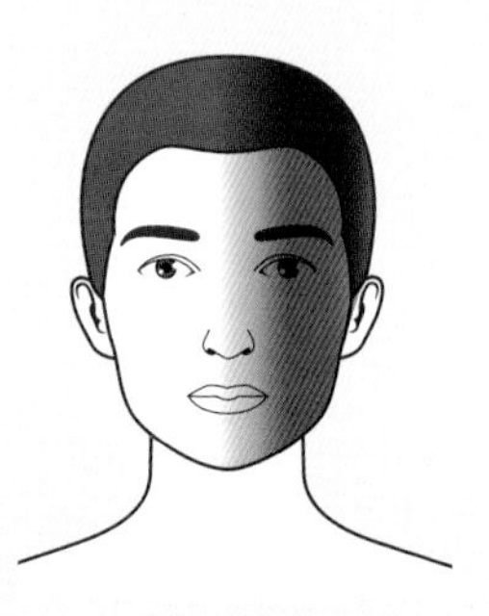

Migraine

Treatment of migraines is more complex than treatment of tension headaches. Milder analgesics are not as effective with migraines but are still the first line of treatment. Due to the complex nature of migraines, very often a combination of medications is used.

side of the brain controls the right side of the body. It is because of the crossing over that occurs in the medulla.

Also, the medulla contains several nuclei (collections of neuron cell bodies). These nuclei help control and regulate many vital body functions. The vasomotor center changes the diameter of blood vessels to adjust blood pressure. The cardiac center regulates the heart rate. Other nuclei in the medulla regulate such things as hiccupping, vomiting, and swallowing.

Before ending our review of the brain stem, there is one other thing I should alert you to—the reticular formation. The word reticular means "net." The reticular formation is a net-like collection of interconnected nuclei that runs through the midbrain, pons, and medulla. A part of the reticular formation is made of sensory axons that run up to the cerebral cortex. This part of the reticular formation is called the reticular activating system (RAS). The RAS stimulates the cerebral cortex during the transition from sleep to wakefulness, helping you to become alert.

TAKING A CLOSER LOOK

Cerebellum

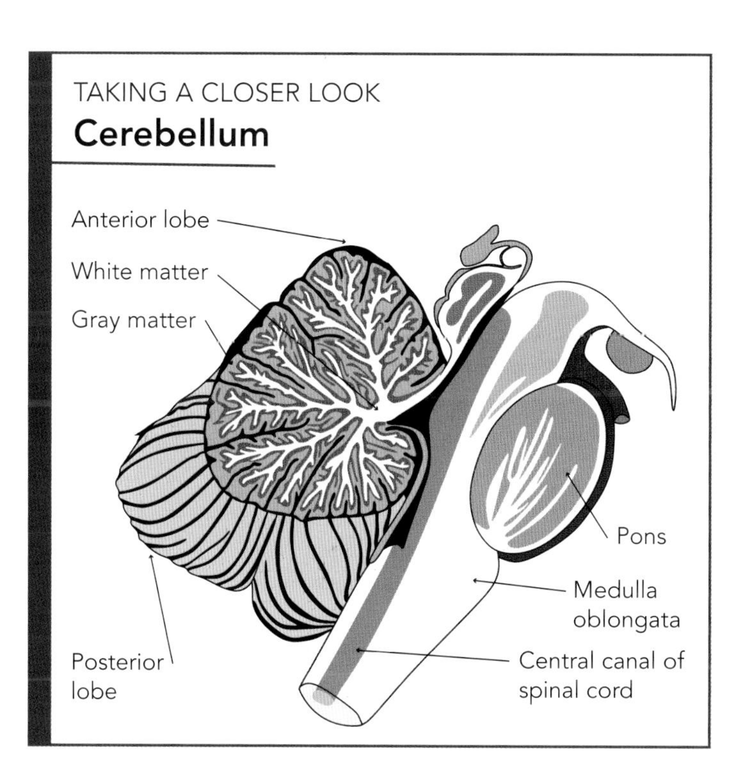

Cerebellum

The final major portion of the brain is the cerebellum. The cerebellum is located posterior to (behind) the medulla. The cerebellum and brainstem are connected by three bundles of white matter called peduncles.

The cerebellum itself consists of a central area called the vermis. Lateral to the vermis are two cerebellar hemispheres. Each hemisphere is composed of three lobes: the anterior lobe, the inferior lobe, and the flocculonodular lobe.

The cerebellum is responsible for making you aware of your position. That is, where the body is, how it is positioned, how fast it is moving. Position sense is called proprioception. The cerebellum receives sensory input from muscles and tendons throughout the body, and processes all these inputs so you will know where all your body parts are located. The cerebellum also helps maintain muscle tone.

Furthermore, the cerebellum helps us maintain our balance. Just think about all the muscle activity that is involved in getting up from a chair. This seemingly simple action takes the coordinated effort of dozens of muscles exerting just the proper amount of force at just the right time. Thank your cerebellum the next time you stand up. Or walk outside on a really windy day. Get the idea?

Blood Supply to the Brain

The brain is the master control center of the body. As such it requires constant high levels of oxygen and nutrients to function correctly and efficiently. Even though the brain makes up only about 2 percent of the body's weight, it requires about 20 percent of the body's oxygen and glucose. Since the brain's metabolic needs are so high, it is not surprising that an interruption of its oxygen supply can have very serious consequences. A loss of oxygen for as little as four minutes can result in permanent brain damage. Let's see how the blood flow to the brain works.

The blood supply to the brain primarily comes from the internal carotid arteries. If you press on your neck gently on either side of your Adam's apple, the pulse you feel is from your internal carotid artery. (Do not check the pulse on both sides at once. Putting pressure on both arteries at once may cause you to lose consciousness.)

After entering the cranium, the right and left internal carotid arteries each branch to form the anterior cerebral arteries and the middle cerebral arteries. As you might guess, the anterior cerebral arteries supply the front part of the brain, and the middle cerebral arteries supply the middle parts of the brain. To help ensure the brain's uninterrupted blood supply, the right and left anterior cerebral arteries are connected by the anterior communicating artery.

Near the base of the neck arteries branches from both the right and left subclavian arteries. (Look back at Volume 2 of the *Wonders of the Human Body* to review the arteries in the neck.) These branches are the vertebral arteries. They ultimately enter the

TAKING A CLOSER LOOK

Blood Supply to the Brain

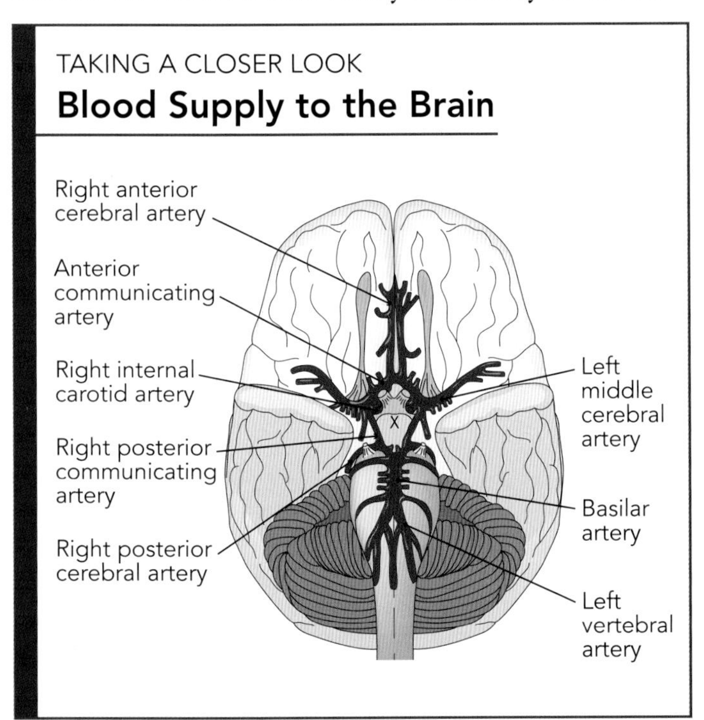

Subclavian Steal Syndrome

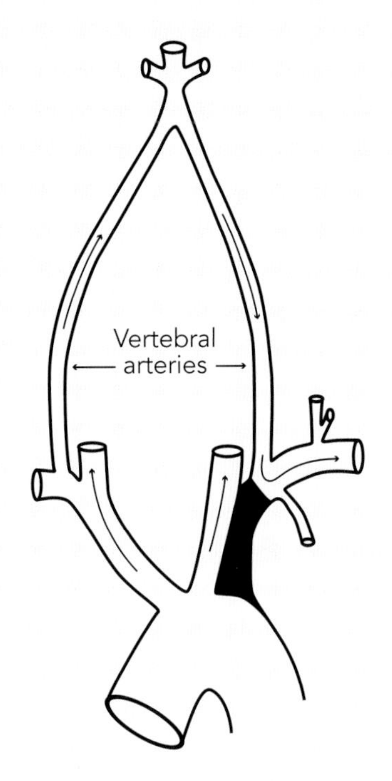

Recall from Volume 2 of the *Wonders of the Human Body* that the subclavian arteries, which are located under the collarbones, supply blood to the arms. You've just learned that the subclavian arteries also supply blood to the brain via the vertebral arteries that branch from them. And since those vertebral arteries merge to form the basilar artery, they are connected to the circle of Willis, giving the vertebral arteries—and hence the subclavian arteries from which they branch—a connection to the brain's blood supply.

Suppose a subclavian artery develops a problem close to its point of origin, becoming so narrow that it cannot keep the arm well supplied with blood. Some of the blood flowing though the circle of Willis can be re-directed away from the brain and flow back down the vertebral artery to the place where the subclavian artery is still open above the blockage. This blood, having bypassed the blockage, helps keep the arm supplied with oxygenated blood.

This condition is called subclavian steal syndrome because blood is "stolen" from the circle of Willis by this subclavian connection. Fortunately, the blood flow to the brain is usually so good that most people with subclavian steal syndrome have no symptoms, though some people experience fainting.

posterior portion of the cranium. After entering the cranium, the two vertebral arteries join to form the basilar artery. After proceeding under the base of the brain, the basilar artery divides into the two posterior cerebral arteries. It will be no surprise that the posterior cerebral arteries supply the posterior parts of the brain. Again, to ensure an uninterrupted blood supply to the brain, each posterior cerebral artery is connected, via a posterior communicating artery, to one of the middle cerebral arteries.

Tucked underneath the brain is an amazing structure called the circle of Willis. It is named after Thomas Willis, the 17th-century physician who discovered it. The circle of Willis is made up of the internal carotid arteries, the anterior cerebral arteries and the anterior communicating artery that connects them, the basilar artery, and its branches, the posterior cerebral arteries, as well as the posterior communicating arteries that connect them to the

Stroke

A stroke occurs when cells in the brain are killed by loss of blood flow. Strokes can result in permanent neurologic damage or even death.

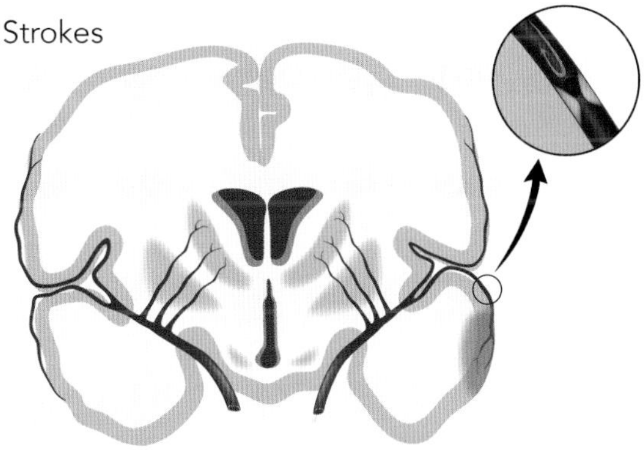

The most common type of stroke is called an ischemic stroke. In this case, blood flow to the brain is interrupted by a blockage in one or more arteries supplying the brain. With no blood flow past the blockage, the nervous tissue supplied by that particular artery is damaged due to lack of oxygen and nutrients. If only a small amount of circulation is interrupted, the resulting damage may be small. On the other hand, if a blockage interrupts a larger circulatory pathway, the damage can be extensive. Blockages can result from the buildup of atherosclerotic plaques in an artery (imagine a clog slowly building up and eventually clogging the kitchen sink) or a blood clot that blocks blood flow.

Another type of stroke is called a hemorrhagic stroke. In this case there is bleeding directly into the space around the brain. Sometimes the bleeding is the result of the rupture of an aneurysm. An aneurysm is a dilated area on the wall of an artery, an area of weakness resembling a thin, weak section of a balloon. At other times an artery in the brain simply ruptures.

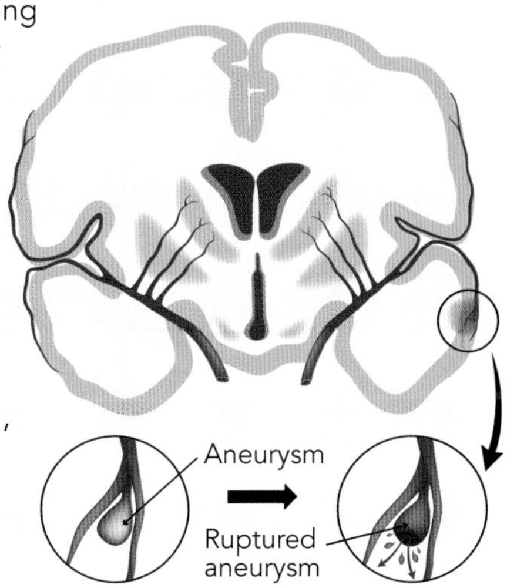

Both types of stroke are very dangerous.

The symptoms of a stroke include sudden loss of function of a part of the body (face, hand, arm, leg, etc.), sudden inability to speak, dizziness, loss of sensation, change in sense of smell or taste, or a sudden change in vision. These are only a few potential symptoms. At times the diagnosis of a stroke can be quite challenging. However, the sooner a stroke victim gets medical attention, the better. Time is of the essence with a stroke.

Rick factors for stroke include high blood pressure, smoking (NEVER, ever, ever, EVER start smoking!!), high cholesterol, diabetes, lack of exercise, and poor nutrition.

middle cerebral arteries. This incredible structure provides much protection for the brain. If a vertebral or carotid artery becomes blocked, circulation to the brain can be maintained (to a certain degree, at least), by blood flow coming from the other arteries making up the circle of Willis.

This arrangement found in the circle of Willis—somewhat circular connections between arteries from both sides—is called collateral circulation. Collateral circulation helps ensure a good supply blood to this important organ.

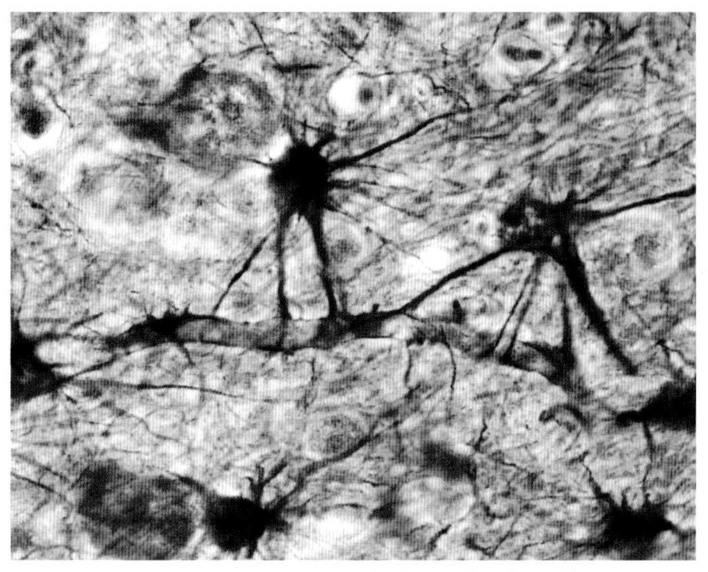

Astrocyte processes in contact with the walls of capillaries.

Blood Brain Barrier

We noted earlier that the brain requires a high percentage of the body's oxygen and glucose to function properly. The brain is also quite sensitive to many substances. To prevent potentially harmful things from coming into contact with brain tissue, there is a blood brain barrier.

Recall that an astrocyte is a type of neuroglia cell. The processes of astrocytes surround the capillaries in the CNS. In doing so, the astrocytes stimulate the endothelial cell lining of the capillary to form tight junctions. Most capillaries in the body have fairly permeable linings that permit many substances to exit the bloodstream and enter the surrounding tissues. Not so in the brain! These tight junctions make the capillaries in the CNS much less permeable than capillaries found elsewhere in the body. This loss of permeability prevents many substances from coming into contact with tissues in the CNS.

From what sort of substances does the brain need to be protected? Various metabolic waste products, proteins, some toxins like the sort that causes

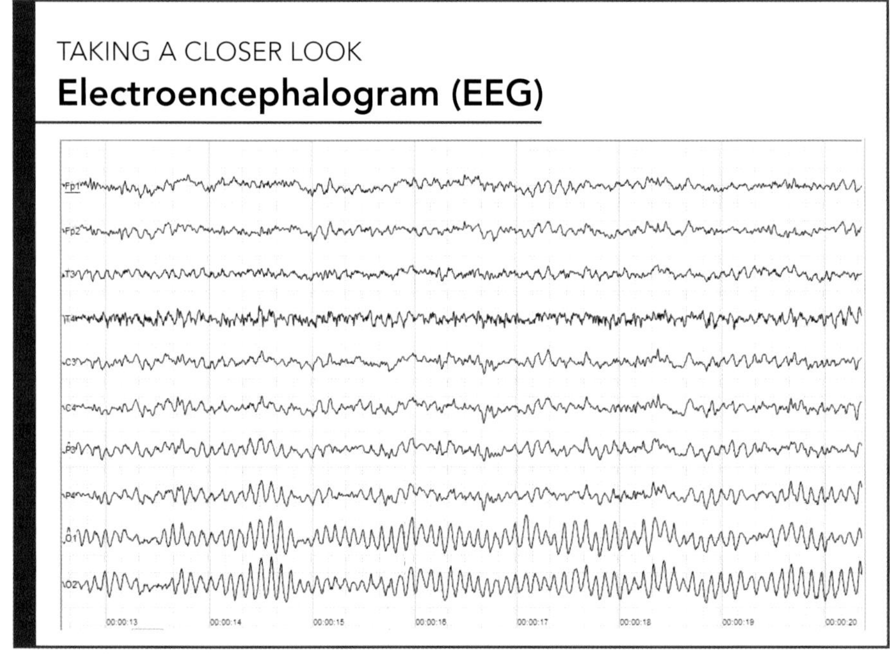

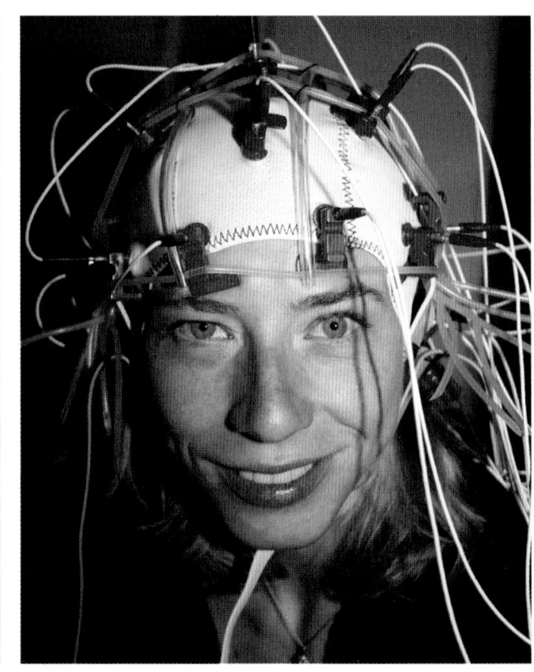

botulism food poisoning, and many types of drugs are stopped by this altered capillary membrane, thus protecting the CNS from their effects. However, oxygen, glucose, and carbon dioxide can easily pass across the blood brain barrier. Fortunately, anesthetics can also pass across the blood brain barrier, making it possible to put people to sleep during surgery (which is fortunate if you ever need your appendix or your wisdom teeth removed).

Brain Waves

The electrical activity that makes your heart beat can be measured on the body's surface with an electrocardiogram (ECG). Similarly, when the brain works, its electrical activity can be measured by means of an electroencephalogram (or EEG, for short). The electrical activity being measured by an EEG is not from action potentials racing down the length of axons. An EEG records the synaptic activity of neurons close to the surface of the brain, those located in the cerebral cortex. The various patterns of electrical activity seen are called brain waves.

There are four basic brain wave patterns, and they can be distinguished by their frequency—how fast they oscillate. Frequency is measured by the number of peaks per second in a waveform. 1 Hz (1 Hertz) means one peak each second. You can see easily distinguish the higher and lower frequencies by examine the waveforms in the illustration.

From lower to highest frequency, the basic brain wave types are:

TAKING A CLOSER LOOK

Brain Waves

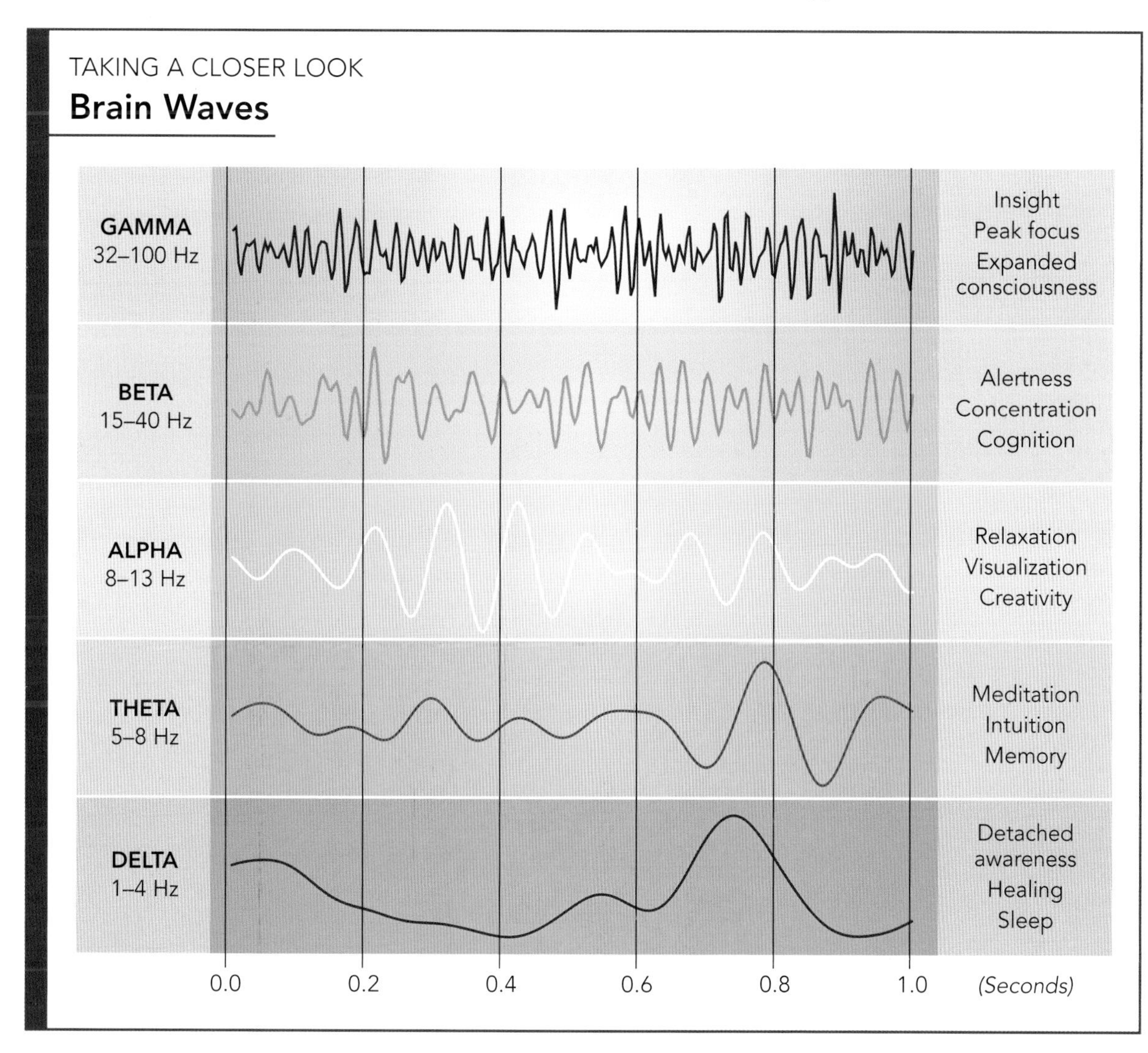

Delta waves — These are usually 1–4 Hz. This pattern is the lowest frequency and occurs in deep sleep. Delta waves in an awake adult usually indicate some type of brain damage.

Theta waves — These are usually 5–8 Hz. In adults, theta waves occur during meditation or drowsiness.

Alpha waves — These are usually 8–13Hz. Alpha waves occur during wakeful but relaxed times.

Beta waves — These are usually 15–40 Hz. Beta waves occur when the mind is active, like when you are concentrating or trying to communicate.

Gamma waves — These are 32–100Hz. This wave type is the least understood of the brain waves. It may be related to consciousness and awareness.

There are, of course, brain wave patterns that are grossly abnormal. The brain wave patterns seen during a seizure would be an example. EEG studies are of primary importance when treating and caring for patients with seizure disorders.

Because the brain is active round the clock even during deep sleep, some degree of brain activity is always detectable by EEG. In the tragic circumstance where there is no detectable brain activity on the EEG, the patient is said to be "brain dead."

Sleep

Everybody sleeps. While it is true that some people sleep better than others, the vast majority of people get at least a few hours of sleep every day. If sleep is something everybody does, exactly what is it?

Sleep is a state in which an individual achieves a degree of unconsciousness from which he or she can be aroused. Although there are several different stages of sleep, here we will explore only the two major types. These are non-rapid eye movement sleep (NREM) and rapid eye movement sleep (REM).

A person passes through stages of non-rapid eye movement sleep as he or she falls deeper and deeper into sleep. The brain wave frequency gets progressively slower with each stage. After 60-90 minutes, the pattern on the EEG changes. This is the beginning of a period of rapid eye movement (REM) sleep. The brain wave pattern during REM sleep is a high frequency pattern, reflecting an increase in neuronal activity. During REM sleep, the body is very relaxed but the eyes move rapidly underneath the eyelids, which of course is how rapid eye movement sleep got its name.

Periods of REM sleep last anywhere from 10 to 60 minutes and recur about every 90 minutes. In the average adult, REM sleep takes up about 25 percent of the total time asleep. Most dreaming occurs during REM sleep. People who do not get enough REM sleep tend to be moody and irritable.

So why do we sleep? What is its ultimate purpose? When it comes right down to it, nobody really knows!

There are many theories about why we sleep. Perhaps sleep is a time that the brain can "take stock of itself." This suggests that the brain needs this time to process all the information from the day that needs to be tucked away and stored or discarded. Others feel that sleep is perhaps a time in which neurons in the brain reset and prepare for the next wake period. Perhaps they do this by removing residual waste products left from the higher activity state of being awake. Maybe sleep helps conserve energy, for we are certainly less active while asleep that while awake. Perhaps sleep gives the rest of the body a time to repair and recover.

In the final analysis, there are a lot of theories about why we sleep, but nobody can say for sure. When I

get 8 hours of sleep each night, I feel better. That's good enough for me!

Learning and Memory

Life sure would be boring if we did not have the ability to learn new things. No one would survive very long if the information in our brain the day we were born was all the information we would ever have. We could not even know how to care for ourselves.

Just consider all the new things you have learned over the past 5 years. Perhaps you have begun to play an instrument or ride a bike. Have you started learning a new language? Or perhaps you have developed an interest in computer programming? You see, learning is not only necessary, it can be fun too!

To learn, we must be exposed to new information. Sometimes this happens because we are being taught something new, like all this incredible stuff you are learning from the *Wonders of the Human Body*! Other times you can learn just by being exposed to situations or circumstances often enough that you eventually learn how something is done. Ride with your dad to the grocery store enough times, and you eventually learn where the store is and how to get there. Check into a cabin with your family on vacation, and you will not only figure out where the kitchen is, but you will soon easily remember where you found it. If your room is upstairs, you very quickly learn that it is easier to go down the stairs than it is to go up the stairs. You do not need to have someone teach you these things. You figure them out by experience.

A vital part of learning is memory. After all, anything you learn is useless to you if you cannot recall it at the proper time. (If you recall your last history test, you will get the idea...or maybe it was all the stuff you could not recall that was the problem.)

The problem is that we do not remember everything we see or experience. Most often we must be exposed to information multiple times before we learn it. Fortunately, God designed our brains to be able to accommodate to new information. When presented with new information, our brains can actually change. Many scientists feel that our ability to remember is the result of synaptic plasticity (now there's your phrase for the day!). This is the ability of synapses to change their strength. This change then results in the encoding of memories.

Amnesia

Amnesia is a loss of memory. Amnesia can result from trauma or severe illness. Certain drugs are also known to cause episodes of amnesia.

There are two primary types of amnesia. The first is called retrograde amnesia. It this case, the person is unable to remember things that occurred before a given event. For instance, after a blow to the head, a severe accident or illness, or a major surgery, a patient may lose to ability to recall some memories from before the event.

The other type of amnesia is anterograde amnesia. Here the person is unable to make new memories after a particular event. In cases like this, it is not unusual for a person to be unable to remember things that happened only minutes before.

Consciousness and the Mind

There are lots of people who believe that the human body is nothing more than a cosmic accident. They would have everyone accept that millions of years ago all the matter in the world simply appeared... out of nothing...from nowhere. Then this matter that appeared from nowhere began to interact with other matter that also appeared from nowhere. As a result of matter interacting with matter, life sprang into being...from lifeless matter.

Then over millions of years the first simple life forms became more and more complex until humans were produced. This process is called evolution. And it is a cosmic fairy tale. You see, if evolution is the correct explanation for our existence, then the human body is nothing more than a chemical accident. Just the result of atoms bumping together over those millions of years.

TAKING A CLOSER LOOK

Spinal Nerves

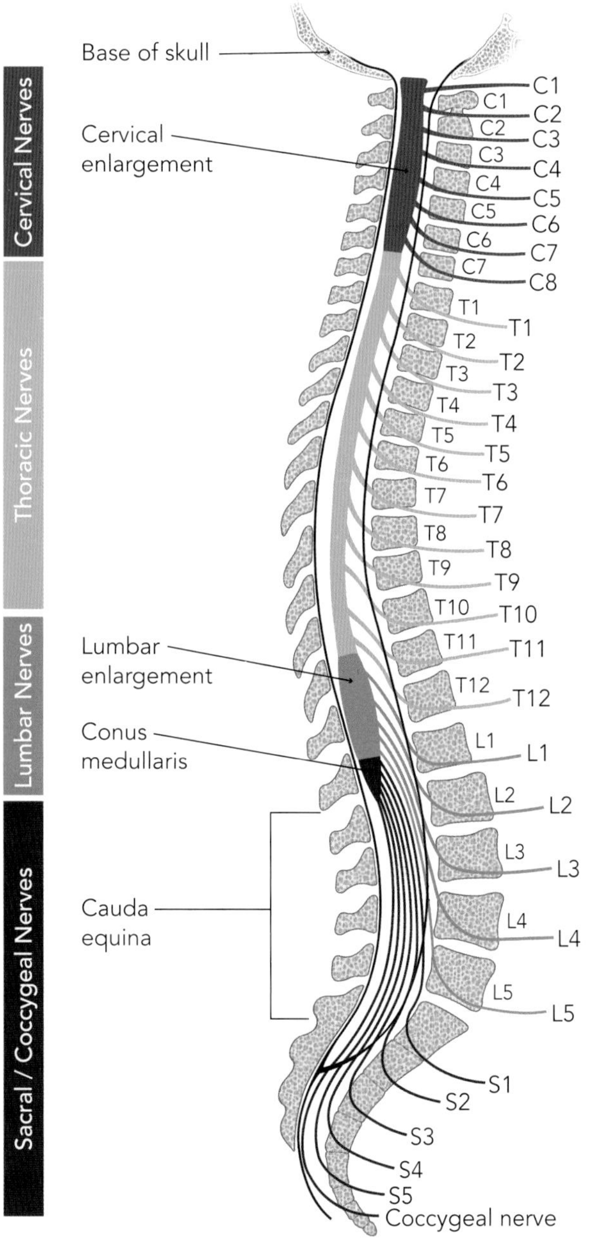

So...how then do evolutionists explain things like consciousness and the human mind? In their view these things must be the result of chemicals interacting with other chemicals. They really have no other explanation. But those people need to understand something.

Professing to be wise, they became fools,

(Romans 1:22)

The human body was designed by the Creator God, and it is more than just a collection of chemicals. Certainly we have a physical body, and it is truly amazing, as you are learning. However, there is much more to being human than that. We not only have a physical body. We also have a spirit and a soul. Humans were created as rational, thinking beings. We are self-aware, thinking creatures. We are able to think abstract thoughts and use language to express them and to communicate them to others.

We have a vast ability to interpret, perceive, and interact with the world around us. We are aware of ourselves and the world around us. This is called consciousness.

Our ability to think and understand is more than just the reactions between neurons and chemicals in the brain. Much more... We are made to think logically, make moral judgements, and use reason to make decisions. A mind is so much more than just chemical reactions and the movement of ions. The mind is a gift from God.

For God has not given us a spirit of fear, but of power and of love and of a sound mind.

(2 Timothy 1:7)

Who has put wisdom in the mind?
Or who has given understanding to the heart?

(Job 38:36)

The Spinal Cord

The spinal cord is the other major part of the central nervous system. It is a very necessary component, for without the spinal cord the brain would be unable to receive sensory information from your body or tell your body what to do. The spinal cord and spinal nerves provide a pathway for sensory information to reach the brain. Equally important, they provide the means for motor output for the brain to reach the body.

The spinal cord starts at the medulla oblongata and extends to roughly the level of the second lumbar vertebral bone in the lower back. In the average adult the spinal cord is about 18 inches in length.

Like the brain, the spinal cord is well protected. It is covered by the same three meningeal layers as the brain, the dura mater, the arachnoid mater, and the pia mater. The meninges, along with the cerebrospinal fluid in the subarachnoid space, cushion the delicate nervous tissue of the spinal cord. Further, the spinal cord is housed in the vertebral canal. The vertebral canal is a long cylinder created by openings in the stacked vertebral bones. The vertebrae surrounding the spinal cord protect it in the same way the cranium provides protection to the brain, even though the vertebral bones move in ways the fused bones of your cranium do not. Thanks to the way the vertebrae are designed, you can bend and twist without damaging your spinal cord or the nerves emerging from it.

Spinal Cord — Gross Anatomy

The spinal cord is roughly cylindrical structure about thickness of a garden hose. As previously described, it extends from the medulla oblongata to the level of the second lumbar vertebra. At its distal end, the spinal cord tapers into a conical-shaped structure known as the conus medullaris.

Emerging sequentially all along the length of the spinal cord are 31 pairs of spinal nerves. The two nerves in each pair exit the spinal cord at the same level, one on the right and one on the left. As nerves

TAKING A CLOSER LOOK

Spinal Cord and Vertebra

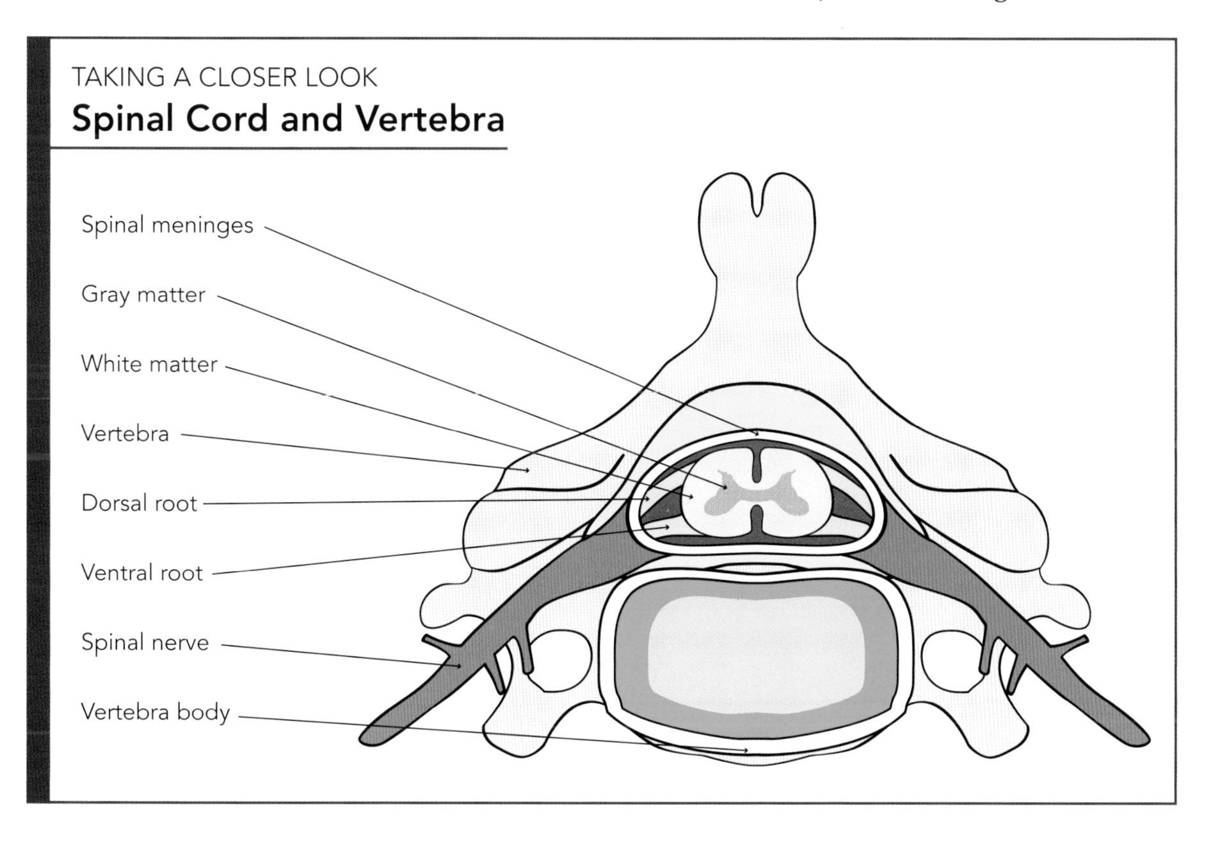

branch from the spinal cord, each leaves the vertebral canal by passing through a window called an intervertebral foramen on their way out to the body.

The name of these windows—intervertebral foramina (plural)—tells us a lot about how they are constructed. Inter- means "between," and the windows are between the vertebrae. Each vertebra in the stacked vertebral column has notches in it, top and bottom. Where a notch on the bottom of one vertebra aligns with a notch on the top of the next one, a window (foramen) is created.

TAKING A CLOSER LOOK

Anatomy of the Spinal Cord

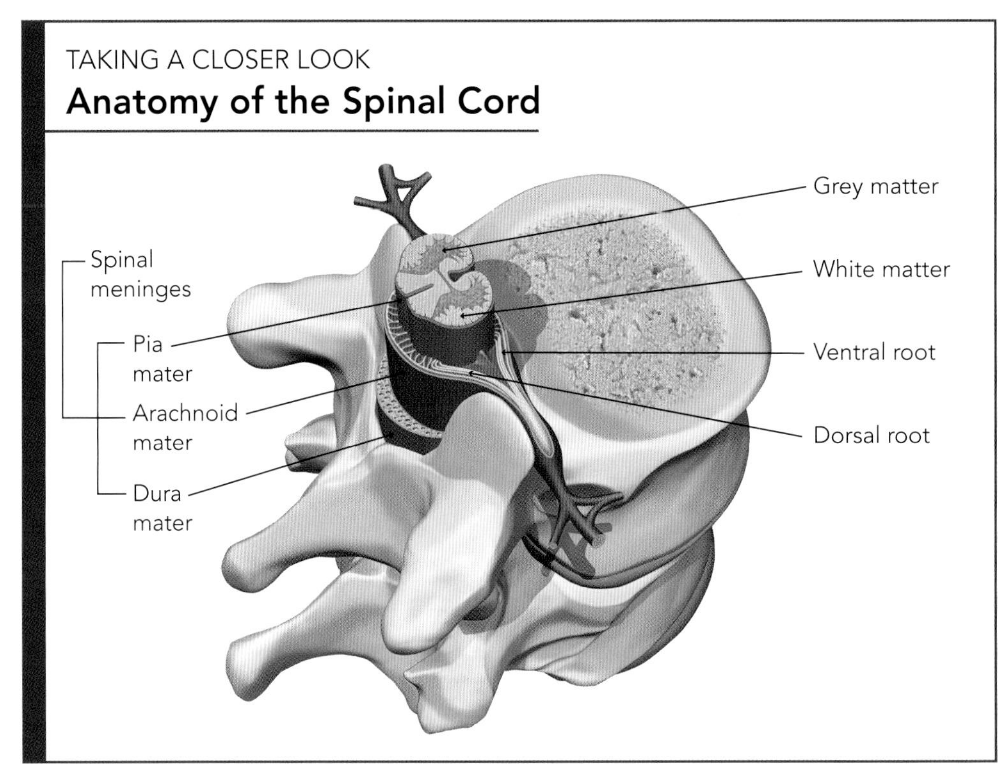

There are eight pairs of cervical nerves (C1-C8), twelve pairs of thoracic nerves (T1-T12), five pairs of lumbar nerves (L1-L5), five pairs of sacral nerves (S1-S5), and one pair of coccygeal nerves (Co1). You might notice something a little odd here. The spinal cord ends at about the second lumbar vertebra, so why are there spinal nerves named for the levels below this?

The naming of the spinal nerves is based on the level of the vertebral column near which they exit, not the level of the spinal cord at which they originate. For cervical nerves one though seven, names are based on the vertebra that is below where they exit. For example, C2 exits above the second cervical vertebra. Then this relationship changes when you get down to C8. Spinal nerve C8 exits below the seventh cervical vertebra and above the first thoracic vertebra. From T1 on down, each spinal nerve is named for the vertebra above its exit.

Here is another puzzler for you. Since the spinal cord ends at about the level of the second lumbar vertebra, where do the nerves from L3 to Co1 come from? That is simple. These remaining spinal nerves emerge from the distal portion of the spinal cord and then travel downward until they exit. This group of spinal nerves extending down from the end of the spinal cord looks like the fibrous tail of a horse. Therefore it is called the cauda equina, which is Latin for "horse's tail"!

Spinal Cord — A Closer Look at the Horns of the Matter

Now that we have seen the big picture, let's take a much closer look at the spinal cord. Each area that gives rise to a spinal nerve is called a spinal cord segment. We will examine a typical spinal cord segment in cross section.

The first thing you likely will notice are the light and dark areas in the main portion of the spinal cord. The dark area is gray matter, and the light area is white matter. Remember, gray matter consists mostly of neuron cell bodies, and white matter consists mainly of axons.

The gray matter projects out in several directions. These projection are called horns, and in cross section they look like a butterfly. The projections directed toward the front are the right and left ventral horns, also known as the anterior horns. Here are found cell bodies of the motor neurons that send signals to "move!" to the skeletal muscles.

The projections to the rear are the right and left dorsal horns (also called the posterior horns). In addition to the cell bodies of interneurons, the dorsal horns contain the axons of sensory neurons. These sensory neurons bring sensory information from the body to the CNS. The cell bodies of these sensory neurons are actually located just outside the spinal cord in a series of dorsal root ganglia. (A ganglion is a collection of nerve cell bodies located outside the CNS.) From each dorsal root ganglion, the sensory nerve axons enter the nearby dorsal horn.

Interneurons are just what their name sounds like they should be: neurons between neurons. As we discussed a while back, interneurons are found in both the brain and spinal cord, forming connections between sensory and motor neurons. Signals from sensory neurons are delivered to interneurons. Interneurons pass those impulses on to the appropriate motor neurons.

In the thoracic and lumbar regions, there are also small horns between the dorsal and ventral horns. These are called lateral horns. Lateral horns contain cell bodies of sympathetic neurons in the autonomic nervous system. These are the neurons that help get the body's organs ready for an emergency.

Spinal Cord —- A Closer Look at White Columns and Roots

Just as the spinal cord's gray matter is divided into horns, so the spinal cord's white matter is divided into columns. If you think of the horns in the cross-sectional view of the spinal cord like the wings of a butterfly, the white matter columns are the in front of, behind, and beside the butterfly's wings. In the rear is the posterior white column. In front is the anterior white column. To each side is a lateral white column.

Each column is made up of tracts. If you recall, bundles of axons in the CNS are called tracts. Each tract is a group of axons headed to the same place (convenient, huh?).

TAKING A CLOSER LOOK

Spinal Cord: Internal Anatomy

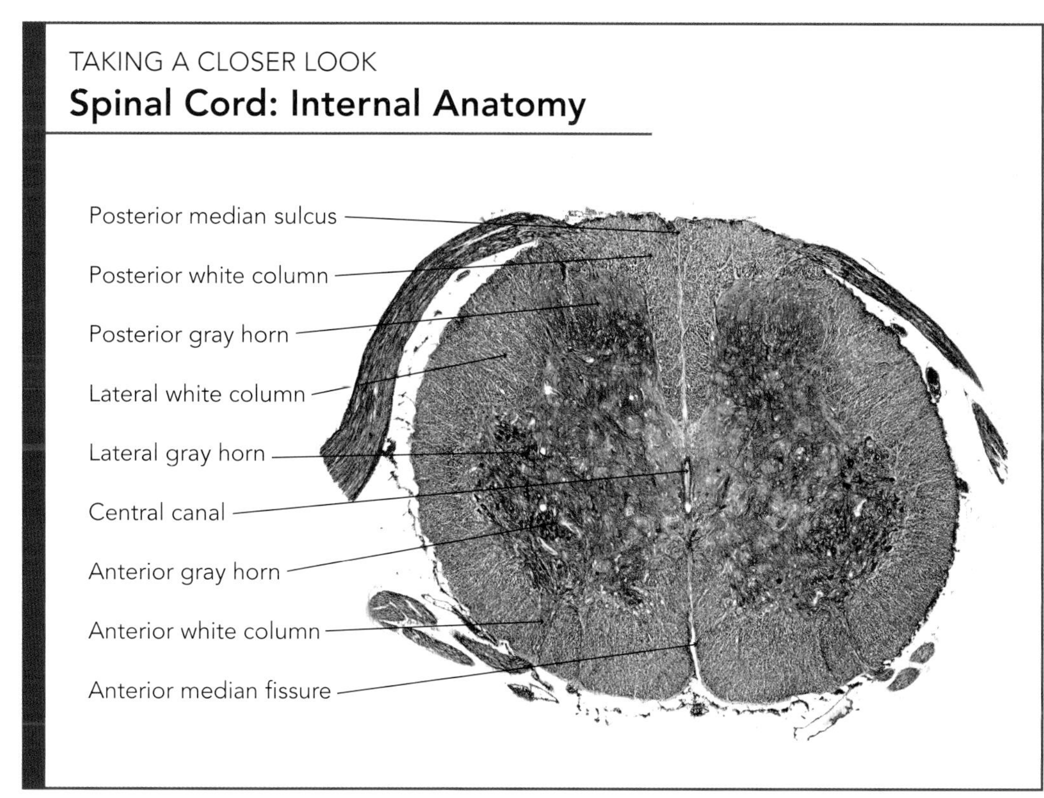

In the spinal cord, the sensory tracts that carry signals to the brain are called ascending tracts. Logical name, right? Sensory signals are sent to the spinal cord and then up (ascending) to the brain. Then the brain integrates all the information. (Remember the three functions of the nervous system?) The brain sends the appropriate signal—the motor output—out to the body. So what do you think the motor tracts are called that take motor output signals down the spinal cord? Yep, they are called descending tracts.

We've seen that the spinal nerve segments have horns and columns, and that the columns consist of lots of tracts. Now let's seen how the spinal cord connects to the spinal nerves that emerge from it. The spinal nerves connect to the spinal cord by means of two bundles of axons, called roots. On each side (right and left) there is a ventral root and a dorsal root. The ventral root contains axons of motor neurons carrying nerve signals from the CNS out to muscle and glands. The dorsal root contains only sensory axons bringing input from sensory

Amyotrophic Lateral Sclerosis

Amyotrophic Lateral Sclerosis (ALS) is a degenerative disease of the nervous system. It is a progressive illness that attacks motor neurons in the ventral horn and the pyramidal tracts. Accompanying this loss of motor neuron function is loss of control of voluntary muscles.

Patients typically present with *fasciculations* (muscle twitching), *atrophy* (loss of muscle mass), and weakness. Over time they often develop difficulty speaking, swallowing, and walking. In later stages of the illness, the patient may be unable to use his arms or legs and lose the ability to swallow or speak. Increasing weakness in the diaphragm and intercostal muscles can result in the patient finally being unable to breathe.

In approximately 90 percent of cases, the cause of ALS is not understood. This is called sporadic ALS. There is another form of ALS, known as familial ALS. This accounts for around 10 percent of cases. This form of ALS is inherited, so there must be a genetic component to the illness in these situations.

At present there is no cure for ALS. Although numerous medications have been tried to aid in slowing the progression of the disease, treatment still consists primarily of supportive care. Most patients die within 3 to 5 years.

Amyotrophic lateral sclerosis also known as Lou Gehrig's disease, so-called after a legendary New York Yankees baseball player who died of the disease in 1941.

Lou Gehrig June 1923

receptors throughout the body. Remember the dorsal root ganglia mentioned above? On each dorsal root is a dorsal root ganglion. The dorsal root ganglion contains the cell bodies of sensory nerves. Sensory input from the body reaches the cell bodies in the dorsal root ganglia, and that input is relayed onward through the dorsal roots to the dorsal horns in the spinal cord. The ventral and dorsal roots merge to form the 31 pairs of spinal nerves.

Tracts in the Spinal Cord

As you might expect, the spinal cord is far more complex than a simple set of two-way streets in which some tracts take signals to the brain and other tracts bring signals from the brain. While the basic concepts remain the same, different types of signals are carried on a wide variety of tracts and processed in different ways.

Thus, the spinal cord is not just a pair of anterior white columns, a pair of lateral white columns, and a pair of posterior white columns. It's more. Much more. Just look at the graphic illustration below. There are many smaller tracts within the bigger areas we defined earlier. This level of complexity is necessary to efficiently integrate all the information the nervous system encounters every second.

Before you feel too overwhelmed with all the fancy names, there is a simple way to work your way through many of them. (And if you decide to go to medical school someday, you will want to remember this simple "trick"!) If you break down the tract names, very often you will know where it goes. For example, the corticospinal tract begins in the cerebral cortex and ends in the spinal cord. Get it? It goes from "cortico-" to "spinal." (The corticospinal tract is one of the pyramidal tracts we talked about earlier.) The "spinothalamic tract" begins in the spinal cord and ends in the thalamus. From "spino-" to "thalamic." So, can you figure out the spinocerebellar tract? Sure you can!

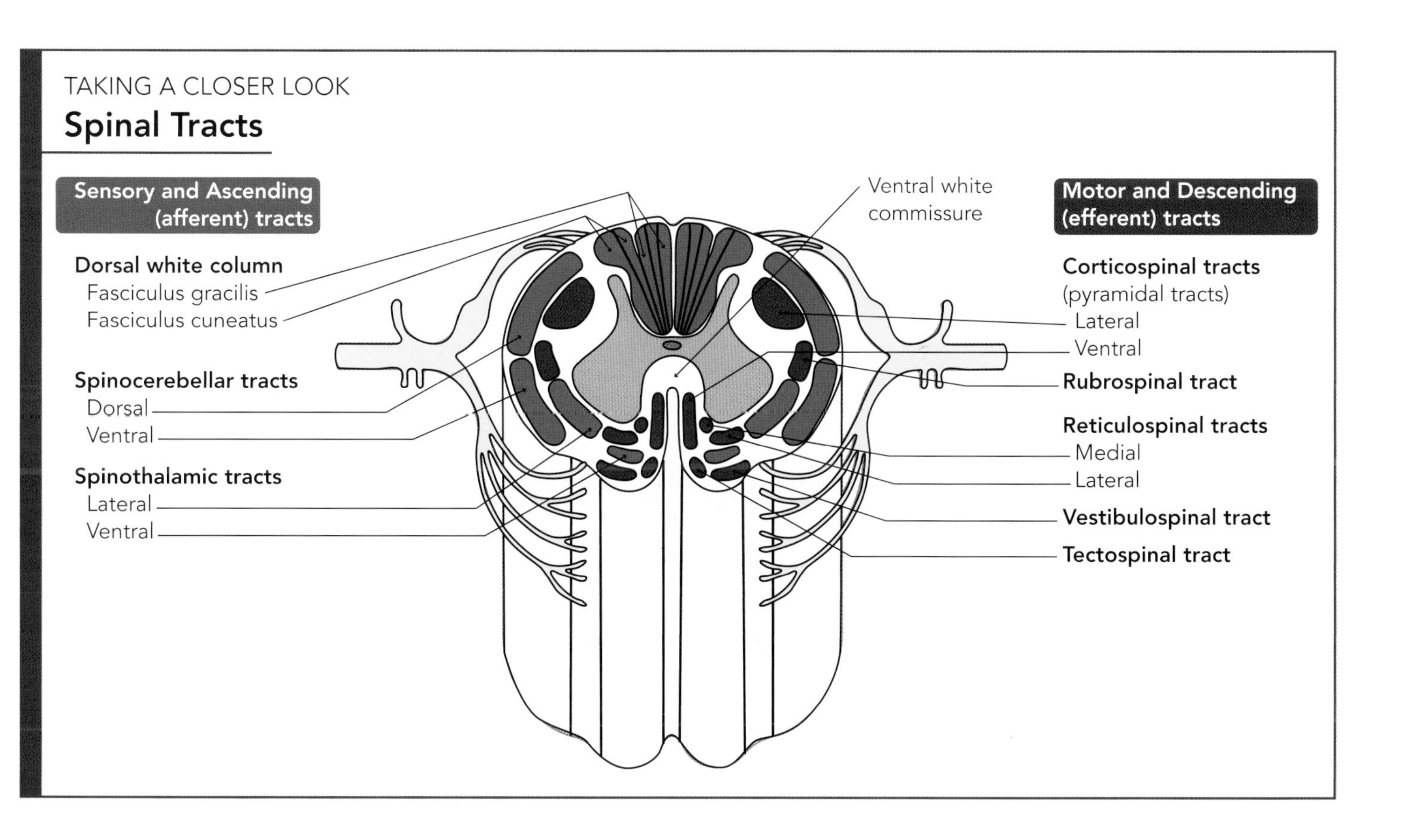

5

THE PERIPHERAL NERVOUS SYSTEM

The peripheral nervous system is the portion of the nervous system outside the brain and spinal cord. It is basically everything else in the nervous system that we have not yet covered. That includes the cranial nerves and the spinal nerves. (We've seen how the spinal nerves begin—we've looked at their roots!—but we haven't seen how they get where they are going.) Cranial nerves emerge from the brain, and spinal nerves emerge from the spinal cord.

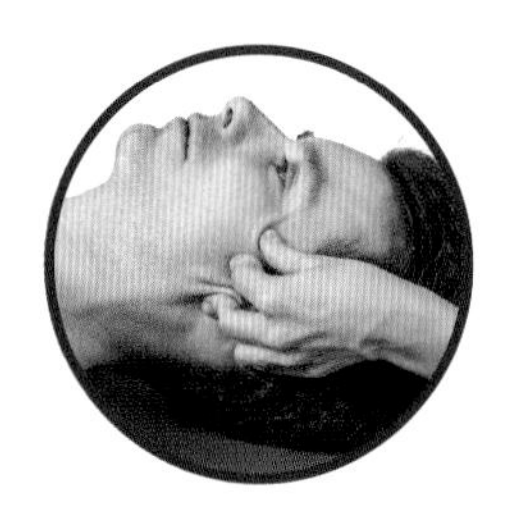

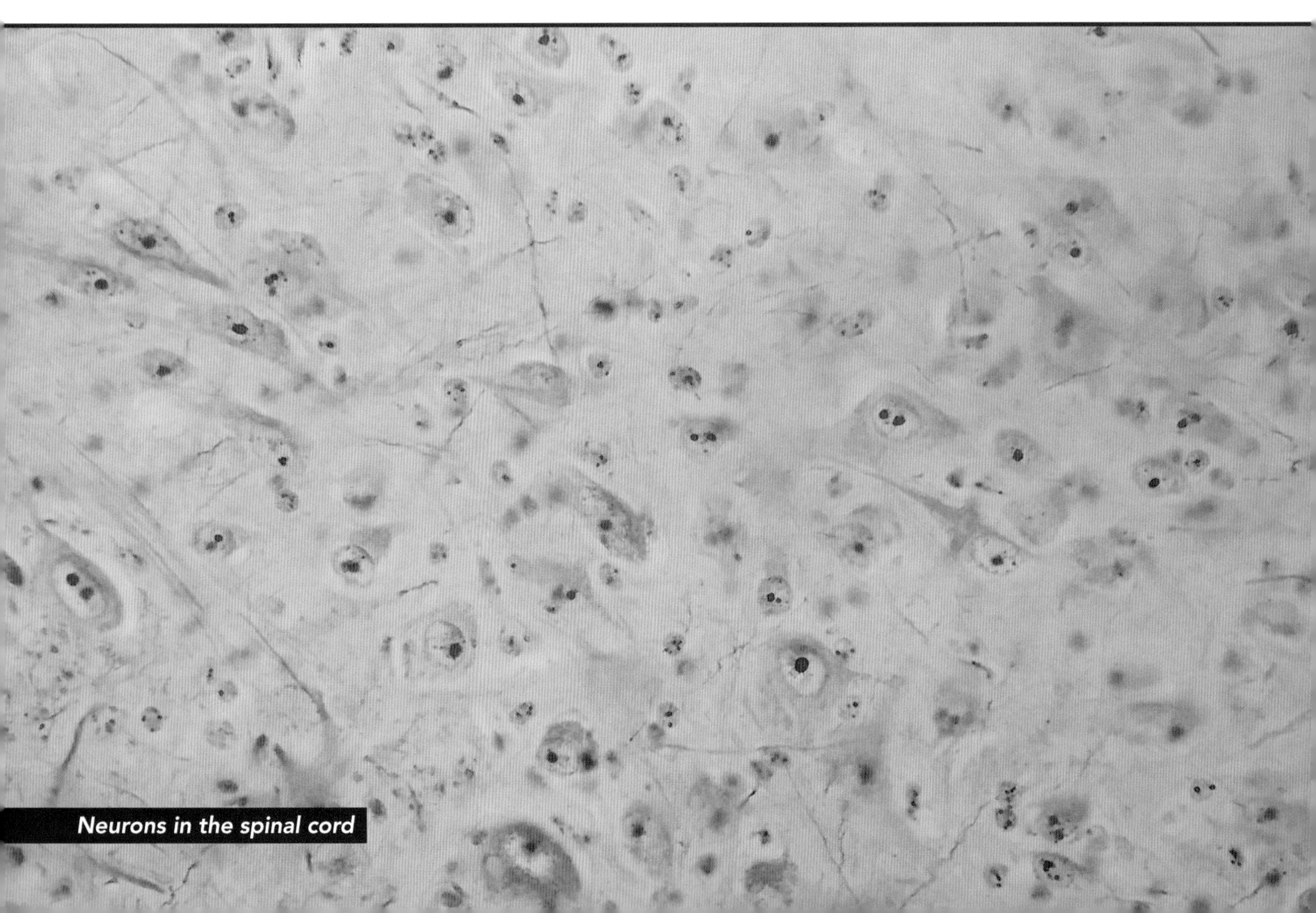

Neurons in the spinal cord

The peripheral nervous system brings sensory signals from the body to the central nervous system, and it takes motor signals out to the various part of the body after the CNS integrates the information. Without the peripheral nervous system, we would not be aware of the world around us. Let's begin by looking at the cranial nerves.

Cranial Nerves

There are 12 pairs of cranial nerves. These nerves emerge directly from the brain and pass through holes (called foramina) in the cranium. Remember, a foramen is a "window," and foramina are "windows." Your cranium has many little windows through which your brain connects to the world using cranial nerves. Even though cranial nerves connect directly to the brain, they are considered a part of the peripheral nervous system.

These nerves are numbered and named, as seen below on the Cranial Nerves image. First, each pair of cranial nerves is numbered based on where it arises from the brain, from front to back. Then, the name of each cranial nerve reflects its function or its path. For instance, the first pair of cranial nerves, I, are associated with smell, so they are is called the "olfactory (I) nerves." The second pair to emerge from the brain, cranial nerves II, are for sight, so they are called the "optic (II) nerves."

We will now present all the cranial nerves in order. Please refer back to the image below to see the nerve locations.

TAKING A CLOSER LOOK

Cranial Nerves

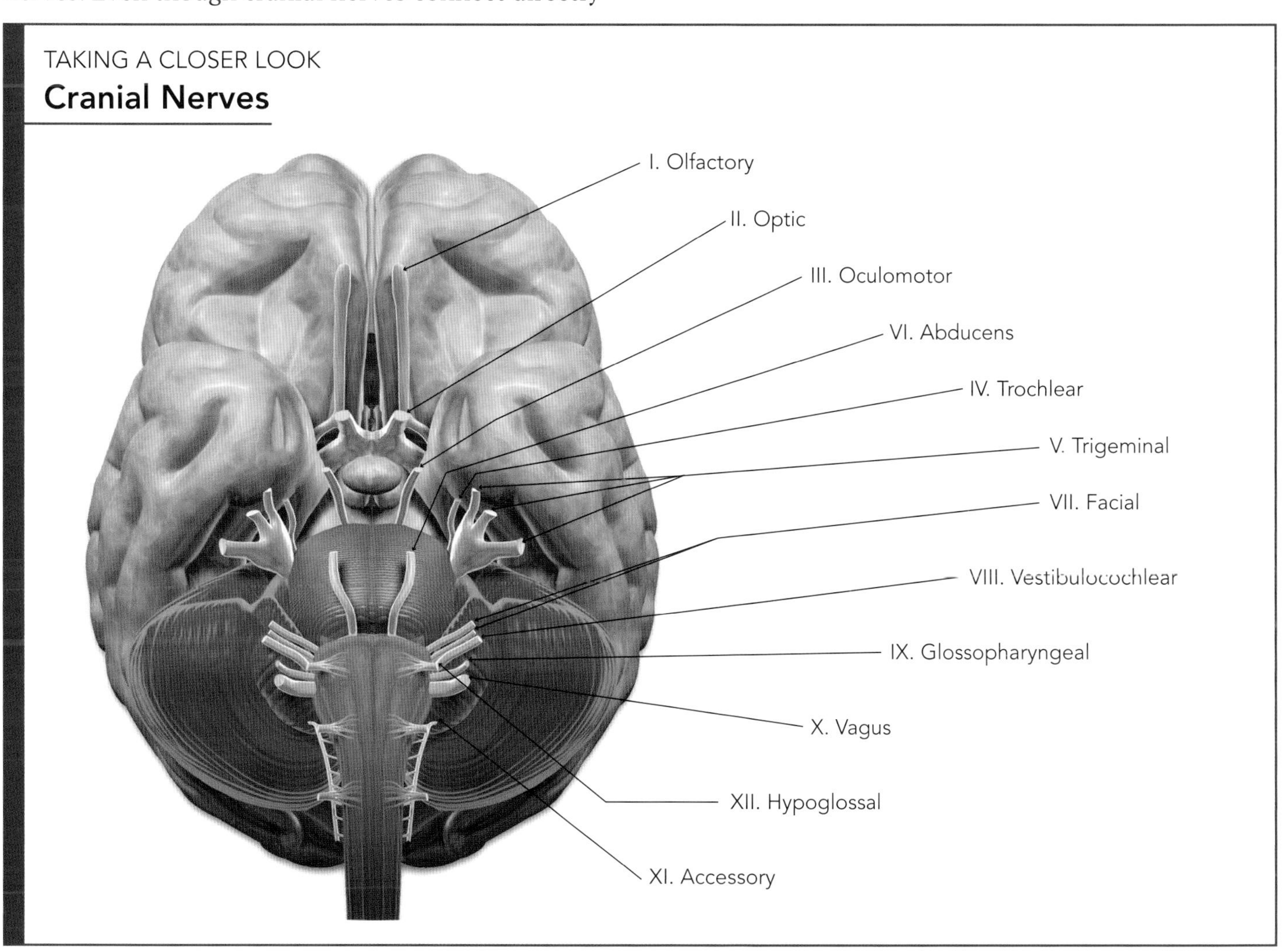

TAKING A CLOSER LOOK

The Olfactory (I) Nerve

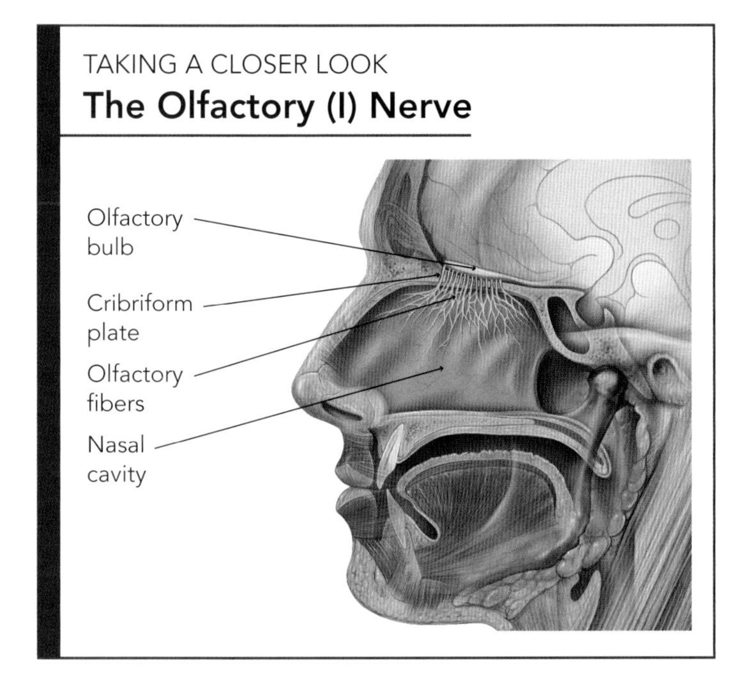

Cranial Nerve I: The olfactory nerve is a sensory nerve. It transmits impulses about smells, or olfaction. Specialized receptors that are tailor made to detect many sorts of molecules in the air are embedded in the lining at the top of the nasal cavity. Axons run from these receptors through a thin bony plate full of holes — the cribriform plate — to the brain's olfactory bulb just above it, and eventually to the olfactory cortex. Cranial nerve I is purely sensory: it detects smells, but it doesn't send your nose any instructions.

Cranial Nerve II: The optic nerve carries nerve impulses for vision. Fibers from each eye's retina coalesce to form an optic nerve, one for the right eye and one for the left. Each optic nerve runs through an opening in the eye's orbital socket to reach the brain.

The optic nerve—cranial nerve II—is purely sensory: it carries information about what you see to your brain, but it does not bring back any instructions for your eyes. Those instructions—instructions telling you which way to point your eyeballs—arrive via three other pairs of cranial nerves—III, IV, and VI. Do you get the idea that what you see and where you look are pretty important? Four of the twelve pairs of cranial nerves are devoted to your eyes!

Trigeminal Neuralgia

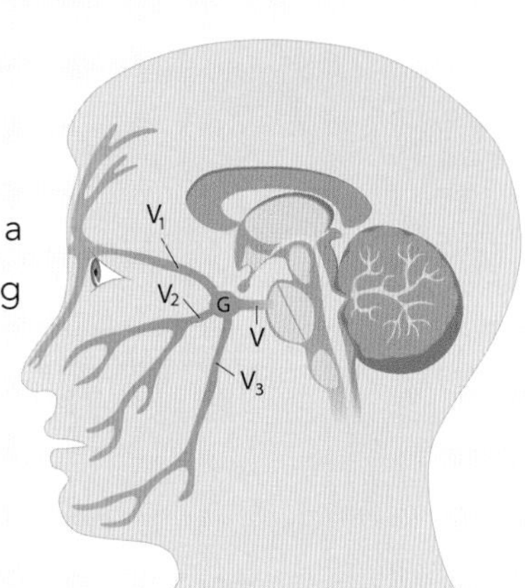

Trigeminal neuralgia is a pain syndrome affecting the trigeminal (V) nerve. It is sometimes called tic douloureux.

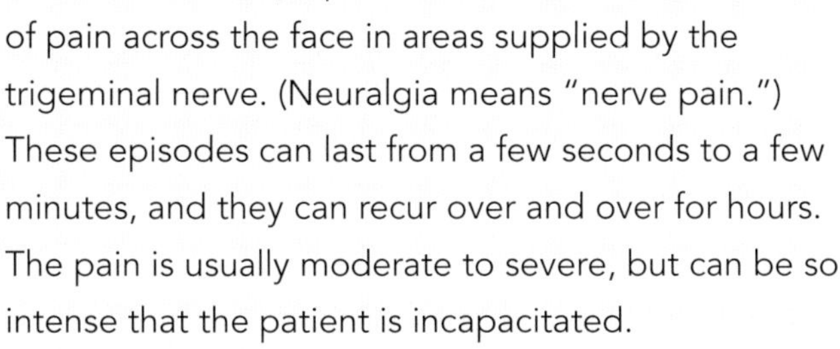

Trigeminal neuralgia is characterized by episodes of pain across the face in areas supplied by the trigeminal nerve. (Neuralgia means "nerve pain.") These episodes can last from a few seconds to a few minutes, and they can recur over and over for hours. The pain is usually moderate to severe, but can be so intense that the patient is incapacitated.

The pain feels like a stabbing or burning sensation. It can affect the scalp, forehead, cheek, nose, teeth and gums. At times the simplest activity or movement precedes an attack. Brushing the teeth, blowing the nose, or even shaving may trigger an episode.

Trigeminal neuralgia may be caused by loss of the myelin sheath surrounding the trigeminal nerve or by pressure from a blood vessel compressing the trigeminal nerve.

Treatment of trigeminal neuralgia is challenging. Controlling the episodes may require very potent medications. Even if they successfully control the neuralgia, such medications can have significant side effects. Pain medications, even very potent ones, are often ineffective in controlling the intense pain. In the most severe cases, surgery may be needed. The goal of the surgery is often to decompress the nerve by putting a cushion between the nerve and local blood vessels. At present several different surgical techniques to control trigeminal neuralgia are being investigated.

Cranial Nerve III: The oculomotor nerve carries motor fibers to four of the six extrinsic eye muscles and the muscles of the upper eyelid. Each eyeball is surrounded by six muscles that enable it to look up, down, left, right, and up and down at angles. These are called extrinsic muscles because they are located outside the eyeball. The oculomotor nerve controls most of the movements of the eyeball.

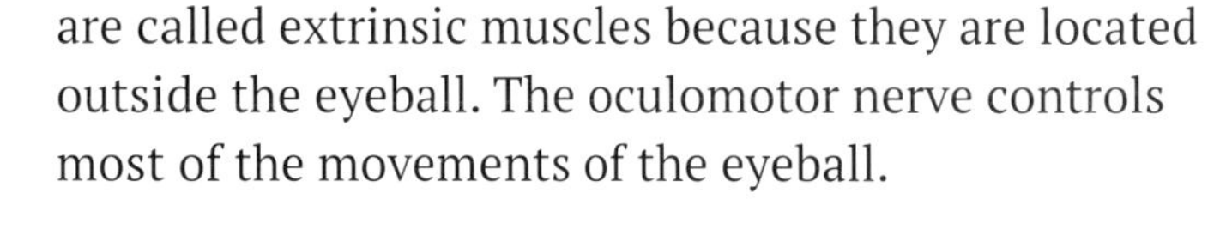

Each eye also has some intrinsic muscles—muscles located inside the eye itself—that adjust the size of the pupil and focus the lens. Autonomic fibers that cause the pupil to constrict are also present in the oculomotor nerve. Cranial nerve III is exclusively a motor nerve; it has no sensory functions.

Cranial Nerve IV: The trochlear nerve controls the movement of the superior oblique muscle of the eye, another of the eye's extrinsic muscles. Cranial nerve IV is a motor nerve.

Cranial Nerve V: The trigeminal nerve provides sensory input from the face. It has three divisions—the ophthalmic, the maxillary, and the mandibular—which supply bring sensory information from the upper, middle, and lower thirds of the face, respectively. The cells bodies from all three divisions are found in the trigeminal ganglion. (Remember, a ganglion is a collection of cell bodies in the peripheral nervous system.)

In addition to its important sensory functions, the trigeminal nerve's mandibular division also has motor fibers that control chewing. Since cranial nerve V has both sensory and motor fibers, it is called mixed nerve.

Cranial Nerve VI: The abducens nerve controls the lateral rectus muscle of the eye, another of the eye's extrinsic muscles. Thanks to this nerve, you can look out to the side. It enables your right eye to look to the right, and your left eye to look to the left. Cranial nerve VI is a motor nerve.

Cranial Nerve VII: The facial nerve supplies motor function to the facial muscles. Thanks to your facial nerves, you can smile, grimace, wrinkle your forehead, and squeeze your eyes tightly shut. The

Bell's Palsy

Bell's palsy is a one-sided paralysis of the muscles in the face. It results from damage to the facial (VII) nerve. Patients with Bell's palsy typically come to the doctor complaining of an inability to smile or frown on one side of the face. Sometimes they cannot close an eyelid, or they may lose taste sensation on the front part of the tongue. Some people with Bell's palsy cannot wrinkle the forehead. The exact symptoms depend on just where the facial nerve is damaged.

The first priority when someone presents with paralysis like this is to be certain the person has Bell's palsy and has not had a cerebrovascular accident (stroke). Usually, a careful examination makes this distinction quickly.

So how does a facial nerve get damaged? Bell's palsy is caused inflammation of the facial nerve, usually due to a viral infection. This inflammation interrupts motor signals to the facial muscles. Most cases of Bell's palsy resolve spontaneously within 3-4 weeks.

facial nerve also carries taste sensation from the front two-thirds of the tongue. Cranial nerve VII is a mixed nerve.

Cranial Nerve VIII: The vestibulocochlear nerve consists of a cochlear branch for hearing and a vestibular branch for balance. You will learn more about this nerve later when we study the ear. Cranial nerve VIII is a purely sensory nerve.

Cranial Nerve IX: The glossopharyngeal nerve contains motor fibers to control swallowing and sensory fibers carrying taste sensation from the rear third of the tongue. It also controls how much you salivate! Cranial nerve IX is obviously a mixed nerve.

Cranial Nerve X: The vagus nerve is the only cranial nerve that extends beyond the head and neck. The vagus nerve provides parasympathetic motor input to the heart, lungs, and abdominal organs. It carries sensory information from the aortic arch near the heart as well as the carotid bodies—collections of sensory receptors near the fork of the carotid arteries in the neck. The brain uses information from these vital avenues of arterial blood flow to help regulate blood pressure and respiration. Cranial nerve X is a mixed nerve.

Cranial Nerve XI: The accessory nerve delivers motor impulses to the sternocleidomastoid and trapezius muscles. These muscles enable you to move your neck and shoulders. Cranial nerve XI is a motor nerve.

Cranial Nerve XII: The hypoglossal nerve provides almost all the motor input to the tongue. Control of the tongue's movement is essential not only for speech but also swallowing. Cranial nerve XII is a motor nerve.

Knowing that an entire pair of cranial nerves is devoted to control of the tongue is a good reminder of the scriptural truth that an uncontrolled tongue can cause a lot of damage.

Even so the tongue is a little member
and boasts great things.
See how great a forest a little fire kindles!
And the tongue is a fire, a world of iniquity...

(James 3:5-6)

Spinal Nerves and Their Distribution

Spinal nerves carry sensory input to the CNS and motor output away from the CNS. These 31 pairs of nerves run the entire length of the spinal cord, exiting at each level. But where do they go after that? There must be some pattern to this. After all, the nervous system controls the entire body, so the nerves must somehow make their way everywhere, right? Yep, right again.

So do the spinal nerves just go directly out to the body structures they supply? Well, yes and no. The spinal nerves in the thoracic (chest) region, T1 to T12, mainly just go to their target areas. Pretty straightforward. However, in the cervical (neck) and

TAKING A CLOSER LOOK

The Cervical Plexus

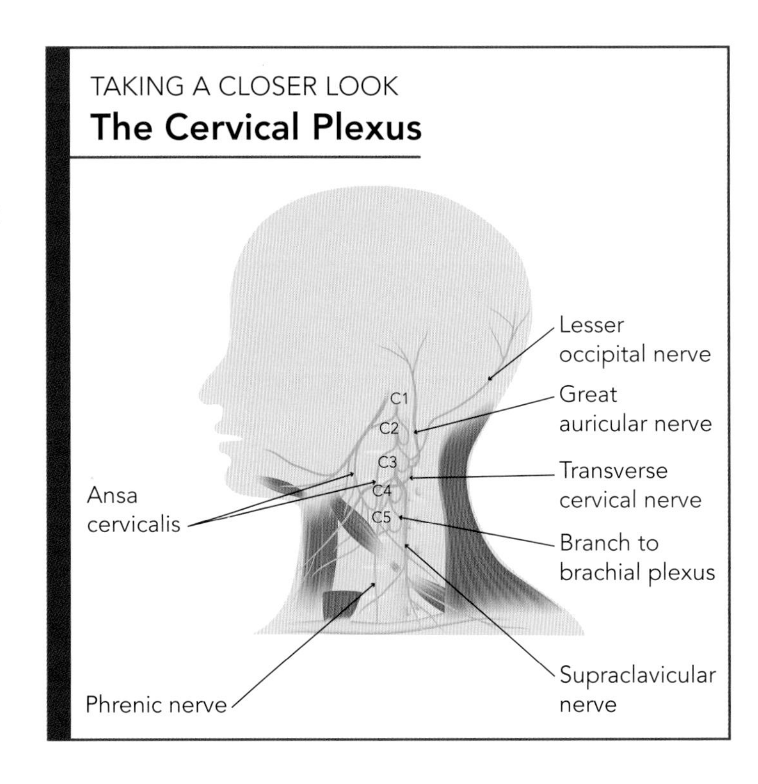

TAKING A CLOSER LOOK

The Brachial Plexus

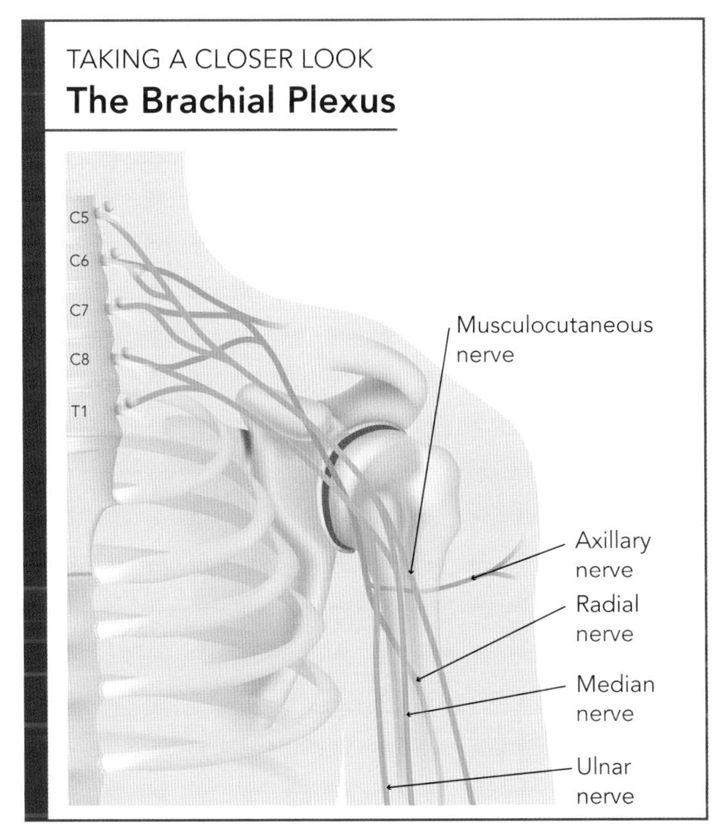

lumbar (lower back) regions the path of the nerves is not so direct.

In these areas when the spinal nerves exit they branch to form complex networks called plexuses. Within a plexus, nerve fibers cross over one another, intermingle, and regroup. Therefore, each branch that ultimately leaves a plexus contains fibers from more than one spinal nerve. The four major plexuses are the cervical, brachial, lumbar, and sacral. The branches that arise from each plexus serve that particular region of the body. We will briefly examine each in turn.

The cervical plexus is formed by branches from the first five cervical nerves (C1 to C5). Through this plexus is carried sensory information from the back of the head, neck, and shoulder. Motor branches supply muscles in the neck. The most important motor nerve coming from the cervical plexus is the phrenic nerve. The phrenic nerve provides motor input to the diaphragm, regularly instructing you to breathe whether you are thinking about it or not.

The brachial plexus is composed of fibers from the C5 to T1 spinal nerves. This plexus is located in the neck and axilla (armpit). The brachial plexus provides the nerve supply to the shoulder and arm. It is incredibly complex, combining and recombining into an array of trunks and cords and branches. The major nerves that extend from the brachial plexus are the axillary nerve, the radial nerve, the median nerve, and the ulnar nerve.

The axillary nerve supplies the deltoid and teres minor muscles. The radial nerve provides motor input to the triceps and the extensor muscles of the forearm. The median nerve supplies muscles of both the thumb's side of your forearm and hand. The ulnar nerve supplies muscles on the medial side (your pinky's side) of the forearm and most of the muscles in the hand. (Be sure to look back at Volume 1 of *Wonders of the Human Body* to see what these muscles do.)

TAKING A CLOSER LOOK

Upper Limb Dermatome

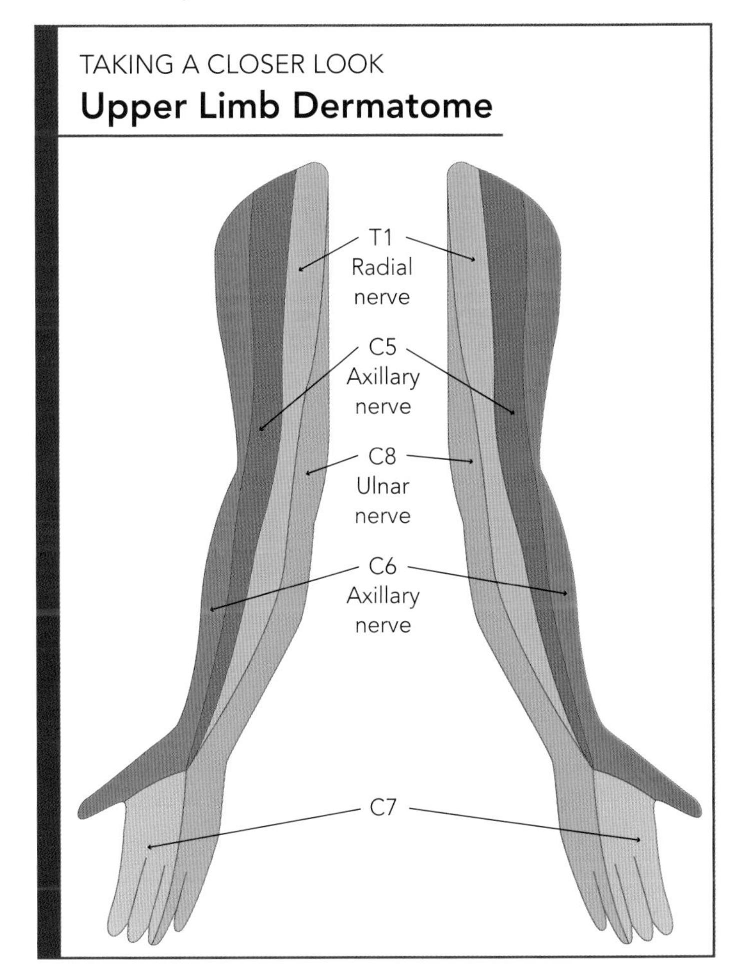

Just an FYI. Have you ever bumped your elbow and hit your "funny bone"? Wasn't really funny, was it? What happened was that something struck your ulnar nerve at a point where it was very close to the surface. You can actually feel your ulnar nerve. There is a small notch (or groove) on the back of the elbow on the pinky side of the arm. If you press too hard in this area, you can feel it tingle. That's your ulnar nerve.

Moving on we come to the lumbar plexus. Thankfully, this plexus is not as complex as the brachial plexus. The lumbar plexus arises from lumbar segments 1 though 4 (L1 to L4). The two large nerves that come from the lumbar plexus are the femoral nerve and the obturator nerve. The femoral nerve is the motor supply to muscles that flex the hip and extend the knee. Among other things, the obturator nerve supplies the adductor muscles of the thigh, the muscles you use to pull your thigh inward.

Finally we come to the sacral plexus. It is composed of fibers from lumbar and sacral spinal nerves (L4 to S4). The nerves from the sacral plexus primarily innervate the buttocks and lower limbs. The primary nerve coming from the sacral plexus is the sciatic nerve. The sciatic nerve supplies the muscles on the back of the thigh, the lower leg, and the foot. It also carries sensory information from the foot and leg.

Phrenic Nerve Injury

As the phrenic nerves control the movement of the diaphragm, injury to these nerves is a very severe situation. If one phrenic nerve is damaged somewhere along its course, perhaps by trauma or a tumor, the diaphragm on that side can be paralyzed. Even more seriously, if the spinal cord is damaged, let's say by a broken neck, above the C3 to C5 level, the diaphragm can be completely paralyzed and the patient will be unable to breathe.

A spinal injury below the C5 level may result in paralysis of the limbs, but in this case the patient would still be able to breathe on his own. This is because the phrenic nerve would be spared in a lower spinal injury.

TAKING A CLOSER LOOK

The Lumbar Plexus

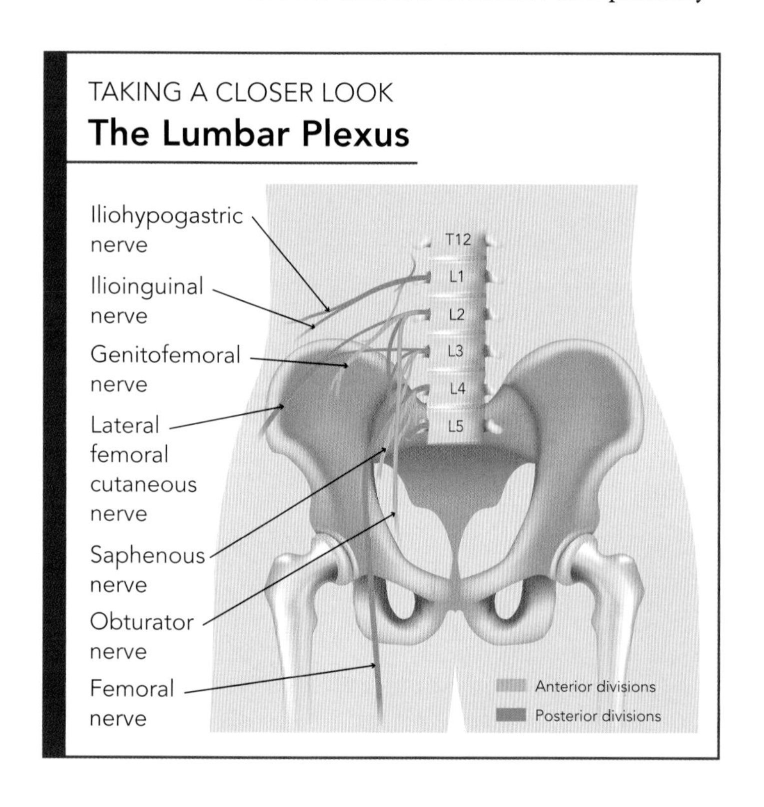

TAKING A CLOSER LOOK

The Sacral Plexus

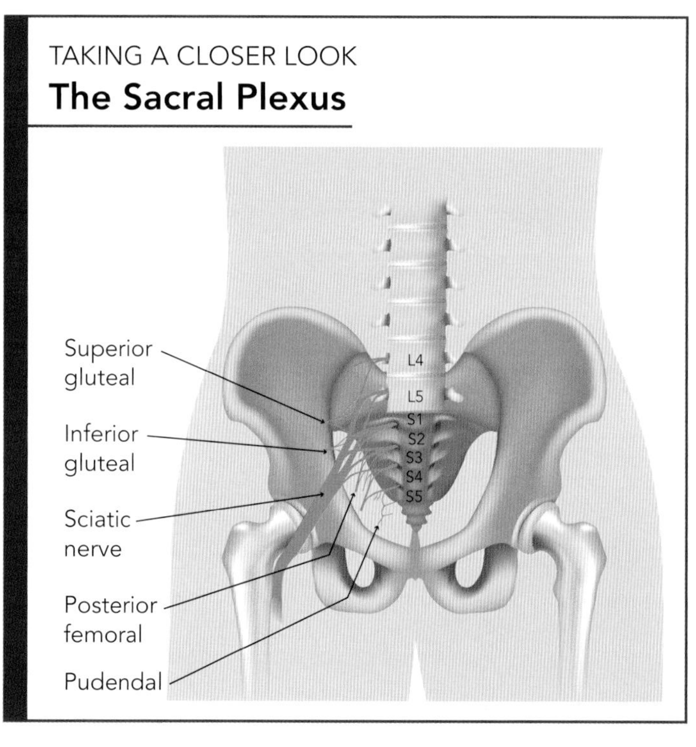

Carpal Tunnel Syndrome

Carpal tunnel syndrome (CTS) can develop when the median nerve is compressed as it runs though the wrist. Carpal tunnel syndrome usually presents as pain and numbness in the thumb, index, and ring fingers—the fingers whose sensation is supplied by the median nerve. The pain can be mild to very severe, even debilitating. As CTS progresses, the pain may even extend up the arm.

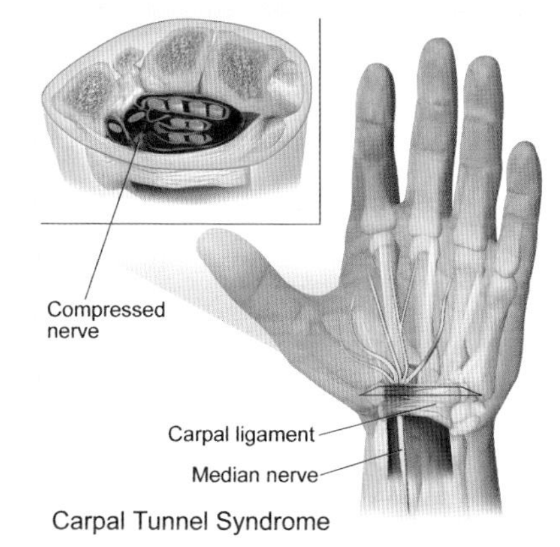

Carpal Tunnel Syndrome

The median nerve passes along the palmar surface of the wrist through an area known as the carpal tunnel. You've probably already guessed (correctly) that carpal means "wrist." This "tunnel" is bordered by wrist bones on one side and a strong, flat, fibrous band (the flexor retinaculum) on the other. Compression of the nerve between the bones and the flexor retinaculum had been cited as the cause of CTS, although there are other theories as well.

Many factors increase the likelihood of developing CTS. These include obesity, arthritis, and diabetes. Although still controversial, the most cited risk factor for CTS is repetitive movement. People who perform repetitive tasks such as typing, using a computer mouse, hammering, or any other activity that puts pressure on the palm side of the wrist are thought to therefore be at higher risk for CTS. This debate will likely continue for some time.

Treatment of CTS includes anti-inflammatory medications, steroids, and the use of wrist splints (particularly during sleep to prevent excessive wrist flexing). In more severe cases, surgery to "release" the flexor retinaculum can be attempted.

Spinal Segments and Dermatomes

If we map the sensory input from your skin to the spinal cord, we see a very interesting pattern. You see, there are sensory receptors in the skin all over the body. These cutaneous receptors send their input to sensory nerves, which transmit these signals to the brain. The regions from which these cutaneous sensory inputs come match the spinal segments. This pattern is very similar in every human body.

Shingles

Shingles, also known as herpes zoster, is a viral disease characterized by painful blisters in localized areas of the body, typically within a particular dermatome. The virus causing this is the varicella zoster virus. The same virus causes chickenpox.

When a bout of chickenpox is over, the varicella zoster virus becomes dormant. It can reside in the dorsal root ganglia of the spinal nerves. If the virus reactivates, it can travel down the sensory nerve fiber and produce blisters on the skin. This is why a characteristic of shingles is patches of painful blisters that follow the distribution of a dermatome on one side of the body.

Treatment of shingles includes both pain medication and antiviral medications. The pain and rash typically resolve in 3-4 weeks. However, some patients have persistent pain in the region affected by the shingles. This can continue long after the rash has resolved. This pain is called postherpetic neuralgia.

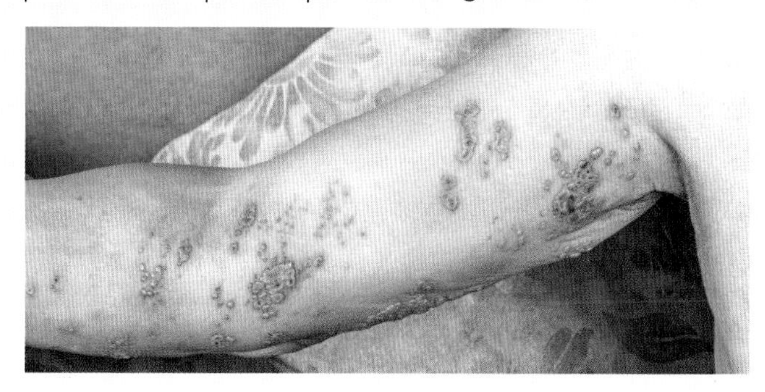

The region of the body that provides sensory input to a particular spinal nerve (or segment) is called a dermatome. Scientists have been able to divide the body into dermatome segments. However, the dermatome segments are not absolute. There is some degree of overlap. For example, the segment mapped to the L2 spinal segment may well send some sensory input to the L1 and L3 segments also. This overlap varies from person to person. Nonetheless, doctors are able to test sensations in the various dermatomes when assessing patients with possible spinal cord damage.

Reflexes

Everyone know what a reflex is, or at least what it does. These are things that the body does automatically, or so it seems anyway. Reflexes are fascinating. They allow the body to perform certain functions or activities without us having to think about them.

A reflex is an automatic motor response triggered by a stimulus. This motor response happens before

TAKING A CLOSER LOOK

Dermatome Map

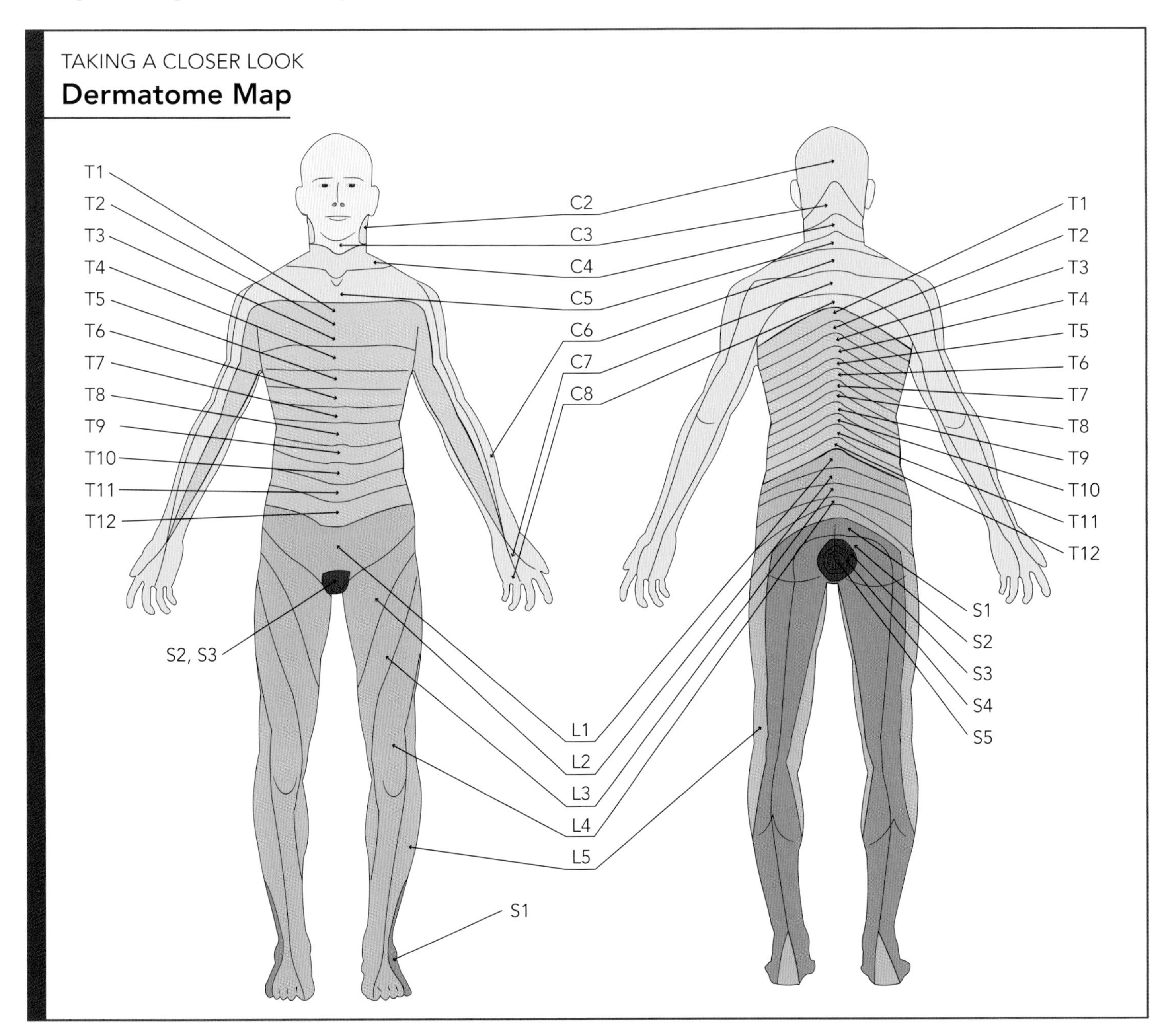

you realize it. Your brain is not involved in the motor output the produces a reflex.

The two primary types of reflexes. One is the somatic reflex. This results in contraction of skeletal muscles. The other is the autonomic reflex. This triggers a response in smooth muscle or glands. With the somatic reflex, you will ultimately be aware of what happened. With an autonomic reflex, you remain unaware of what happened.

Here is the classic example of a somatic reflex. You are walking through the kitchen. You casually place you hand on the top of the stove, not knowing it is very hot. Practically as soon as your hand hits the stove top, it pulls back to get away from the heat. By the time you yell out, you realize that your hand is in the air. Before your brain told you, "Get your hand off the stove; it's really hot!" your hand was off the stove.

Let's examine this typical somatic reflex in detail.

When you placed your hand on the hot stove, sensory receptors in the skin of the hand sensed the heat. These receptors produced a signal that was then transmitted to a sensory neuron. This neuron then carried the impulse through the dorsal root of the spinal nerve, ending in the dorsal horn of the gray matter in the spinal cord.

In the dorsal horn, the sensory neuron can synapse in either of two ways. It may synapse directly with a motor neuron. This involves a single synapse, and it is known as a monosynaptic reflex. (Note that in this case, the integration function of the nervous system

TAKING A CLOSER LOOK

Somatic and Autonomic Reflexes

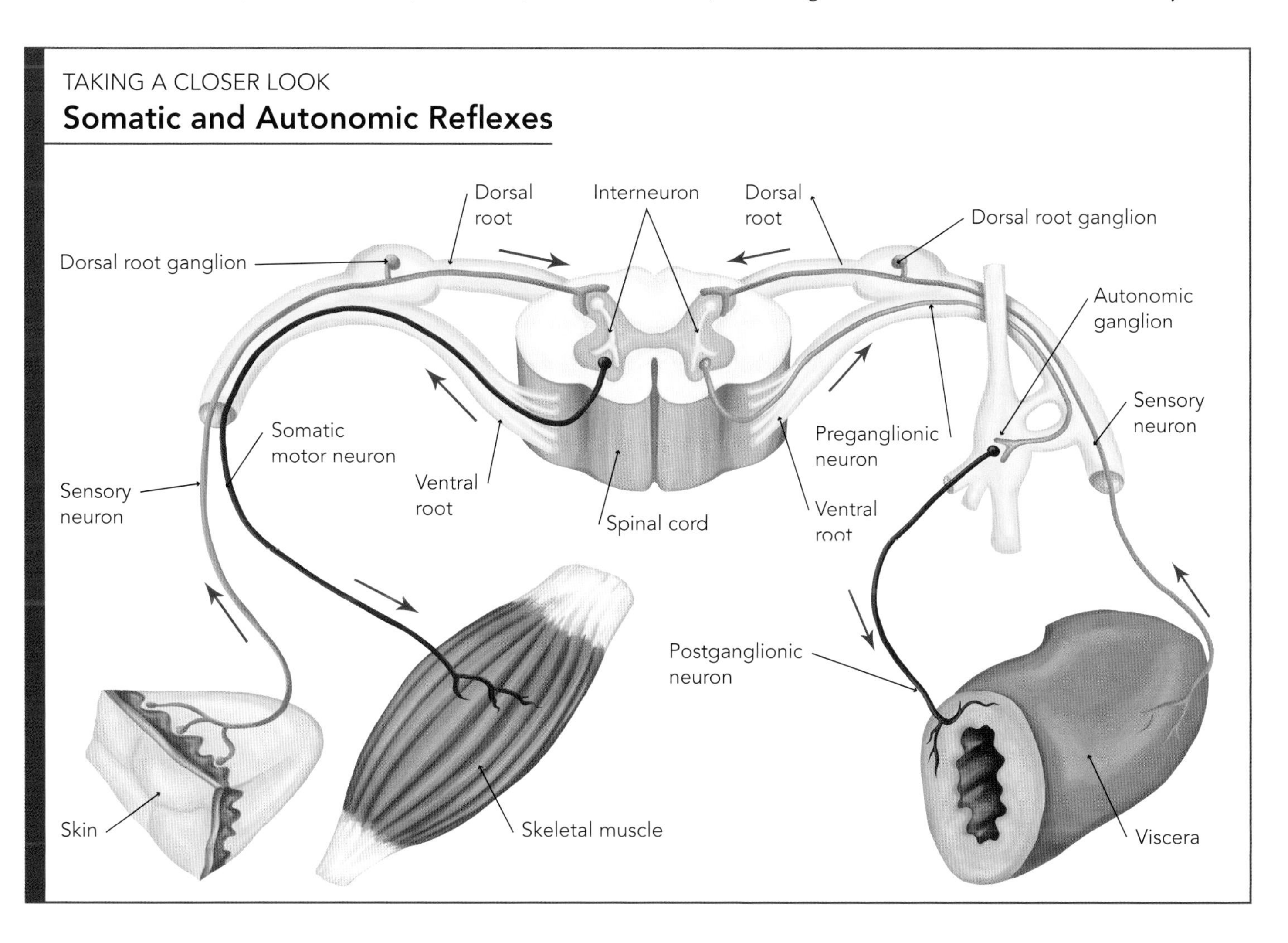

is this single synapse.) In other cases, the sensory neuron synapses with one or more interneurons. This is a polysynaptic reflex.

A nerve signal is then sent out to skeletal muscles by way of motor fibers in the ventral root. The signal is received by the appropriate muscles, prompting you to take you hand off the hot stove! All before you brain even knew.

Of course your brain does find out. Other pathways and connections inform your brain that your hand is now red and burning. This information is processed in the brain and hopefully stored in your long-term memory. Then the next time you are walking though the kitchen, you will ask yourself, "I wonder if that stove is hot?" before you put your hand on it. Learning is a wonderful thing.

Broken down into its basic components, this reflex arc consists of: a sensory receptor, a sensory nerve, an integration center in the spinal cord, a motor nerve, and an effector (the skeletal muscle that withdraws your hand).

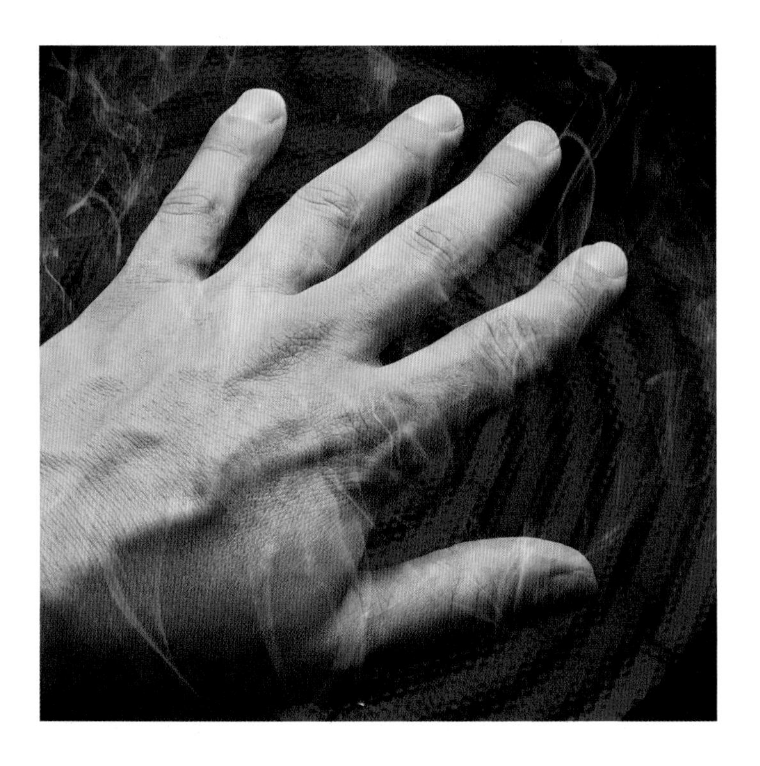

Sensory Receptors

When the wind blows on your face, you touch that hot stove top (remember?), you detect a pebble in your shoe, or you smell brownies baking in the oven, you instantly recognize and identify what you are feeling. How does your body detect and process these things? Your sensory receptors must first detect the stimulus—the wind, the heat, the pressure, the smell.

Sensory receptors in the nervous system allow us to be aware of changes in our environment. These changes are called stimuli, and they occur all around us, to one degree or another, practically all the time. A sensory receptor "receives" a stimulus. This is turn triggers a nerve impulse that can be delivered to the central nervous system.

There are many ways to classify sensory receptors. Each classification scheme has its strengths and weaknesses. We will group sensory receptors according to the type of stimulus they detect.

A thermoreceptor is triggered by changes in temperature. There are—not surprisingly—two types of thermoreceptors. One senses warm, and the other senses cold. These receptors are located near the skin surface. They are activated by moderate degrees of warm and cold. Extremes of warm and cold activate special receptors called nociceptors, which respond to more painful stimuli.

An osmoreceptor senses changes in osmotic pressure. This type of receptor is found mainly in the hypothalamus. Osmoreceptors in the hypothalamus detect changes in the concentration of substances dissolved in the blood because of the pressure created on them when concentrations change.

When such changes in the blood are detected, the hypothalamus can send a message to the posterior pituitary gland nearby, instructing it to increase or decrease its secretion of a hormone called vasopressin. Vasopressin, in turn, lets the kidneys know how much water to hold onto or eliminate from the body. You can see that this is an important homeostatic mechanism without which the concentrations of dissolved substances in the blood could soon change more than is good for us.

A photoreceptor responds to changes in light. Receptors like these are found in the retina of the eye.

Chemoreceptors are triggered by exposure of certain chemicals. Olfactory receptors are a good example. This receptor can detect chemicals in the air. Different chemicals are interpreted by the nervous system as different smells. Taste buds also work by means of chemoreceptors.

Mechanoreceptors sense mechanical force such as stretch, touch, pressure, or vibration. Tactile sensations like touch, vibration, itching, and tickling are mediated by mechanoreceptors.

Proprioceptors sense position in relation to other parts of the body. Muscle spindles within your skeletal muscles are proprioceptors.

A nociceptor are the receptors for painful and potentially damaging stimuli. Things that might damage tissue can trigger nociceptors. Extremes of heat or cold, excessive pressure, and even tissue irritation due to chemical exposure are sensed by nociceptors.

Sensory receptors can be found in skeletal muscle, skin, joints, and visceral organs. (Visceral organs are organs in the chest or abdomen.) By means of these sensory receptors, the CNS can keep up with everything that is happening in and to the body.

The Autonomic Nervous System

Up to this point, our study has focused on the somatic nervous system. Here sensory inputs that we are conscious of such as touch, temperature, pain, taste, and sight are received by the CNS. Then motor output signals are sent to skeletal muscle. When the skeletal muscle is stimulated, its membrane is excited, and as a result it contracts. Therefore, the motor output from the somatic nervous system can only stimulate. We will see that the autonomic nervous system, in contrast, can either "dial up" (stimulate) or "dial down" (inhibit) the targets it affects.

The autonomic nervous system (ANS) receives sensory input from sensory receptors in visceral organs—like the heart, stomach, and intestines—and blood vessels. Baroreceptors, chemoreceptors, and mechanoreceptors monitor blood pressure, heart rate, and respiration. We are not generally aware of these sensory inputs. Our autonomic nervous system monitors all these things day and night. When these inputs are integrated, the motor output from the autonomic nervous system goes to smooth muscle, cardiac muscle, and glands. Again these targets—the effectors—that autonomic motor outputs affect are not under our conscious control.

Remember we said that the autonomic nervous system can either stimulate or inhibit? Well there are two divisions to the autonomic nervous system, and they tend to have opposite effects on any given target. These divisions of the ANS are called sympathetic and parasympathetic. Although many organs receive input from both divisions of the ANS, there are exceptions. There are several effectors that receive only sympathetic innervation.

Sympathetic and parasympathetic signals have opposite effects on a given organ. In any given situation, one will stimulate and the other inhibit. One division will excite; the other will depress. Therefore, in contrast to the somatic nervous system which only excites effectors (namely, skeletal muscles, right?), the ANS has the capability to both excite and inhibit.

The ANS plays a vital role in maintaining homeostasis. You recall that homeostasis is the body's ability to use many interacting mechanisms to maintain balance or "equilibrium" among its many systems. The two divisions of the ANS make it possible to adjust many vitally important variables upward or downward to keep them in the narrow range that is safe for us.

Being able to detect changes in the body's internal environment and to process that information is one thing. But to then be able to immediately either excite or inhibit the effector organs to maintain internal balance, or homeostasis, is quite another thing entirely. Quite a brilliant system.... One that we need to explore further.

Anatomy of the Autonomic Nervous System

You recall that the autonomic nervous system is part of the peripheral nervous system. And you recall that a ganglion (plural, ganglia) is a collection of neuron cell bodies in the peripheral nervous system. Well, ganglia are an important part of the ANS.

TAKING A CLOSER LOOK

Sympathetic Nerves in the ANS

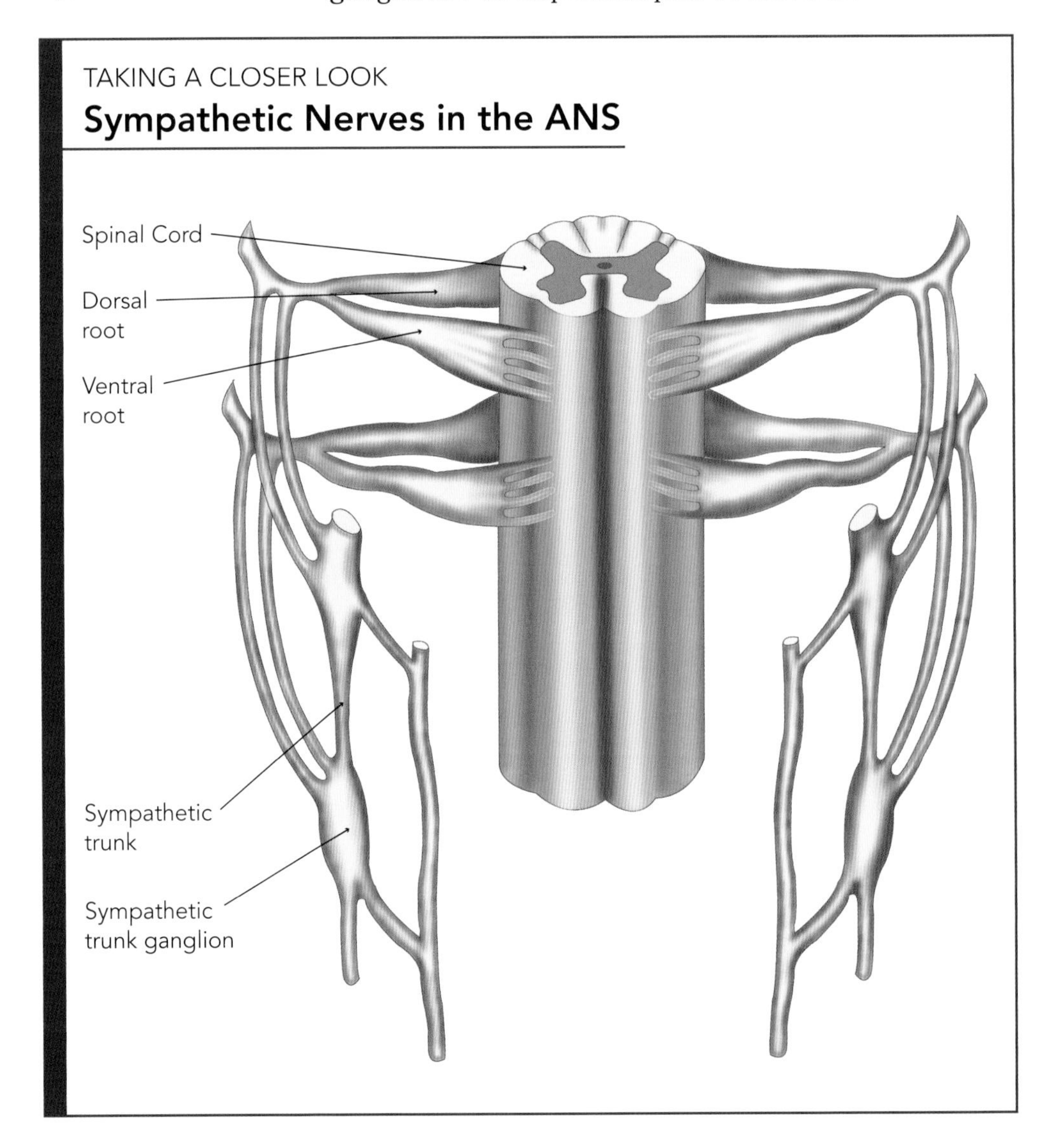

Each motor pathway in the ANS is composed of two neurons, the preganglionic neuron and the postganglionic neuron. The cell body of the preganglionic neuron is in the brain or the gray matter of the spinal cord. Its axon exits the CNS by way of a spinal nerve or a cranial nerve. That axon goes to a ganglion, where it synapses with the dendrites of a postganglionic neuron.

The dendrites and cell bodies of postganglionic neurons reside in ganglia. In the ganglion, the postganglionic cell can synapse with many preganglionic nerve cells.

In the sympathetic division of the ANS, preganglionic neuron cell bodies are located in the lateral horns of the 12 thoracic segments and the first two lumbar segments. These neurons are relatively short, sending their axons to ganglia located very close to the vertebral column in the sympathetic trunk. A chain of these ganglia run parallel to the vertebral column, one on each side.

Another chain of sympathetic ganglia are found in front of the vertebral column and are called the prevertebral ganglia. Preganglionic neurons synapse with postganglionic neurons in the sympathetic trunk and in prevertebral ganglia. Because their preganglionic cell bodies are located from T1 to L2, the sympathetic division is also called the thoracolumbar division.

In the parasympathetic division, cell bodies are found in brain in the nuclei of four cranial nerves (III, VII, IX, and X) and in the lateral horns of the gray matter in sacral segments two through four (S2 to S4). With preganglionic neuronal cell bodies located in the brain or sacral spinal segments, the parasympathetic division is also called the craniosacral division. The axons of preganglionic neurons in the parasympathetic division tend to be much longer than those in the sympathetic division. They synapse with their postgangionic neurons in ganglia (called terminal ganglia) away from the spinal cord and nearer the visceral organs.

TAKING A CLOSER LOOK

Synapse in Trunk Ganglion at the Same Level

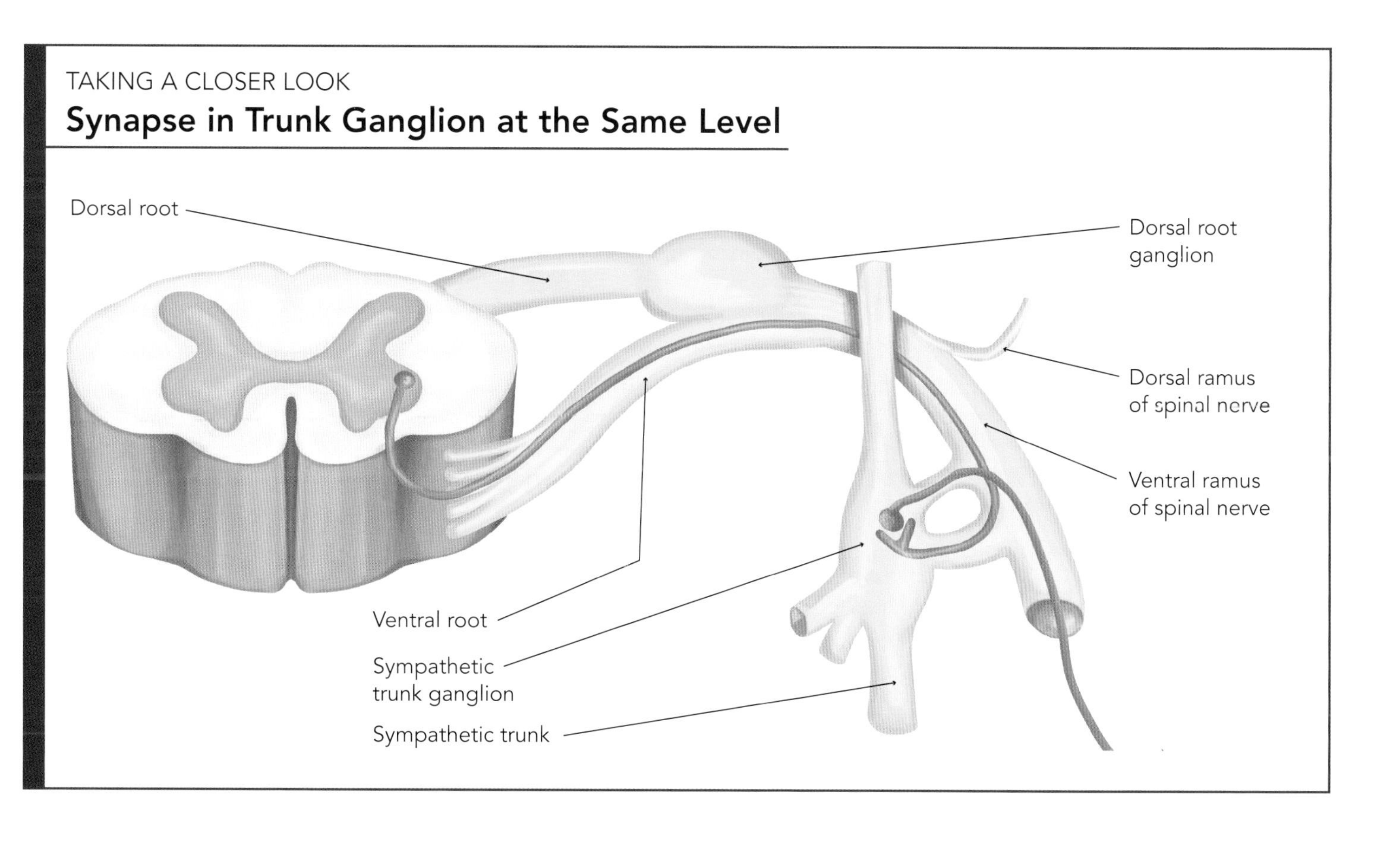

Autonomic Plexuses

Neurons of the somatic motor system formed networks of fibers called plexuses. The same cam be said for ANS. In the thorax, abdomen, and pelvis, there are multiple autonomic plexuses composed of axons from both sympathetic and parasympathetic neurons. These frequently are found near larger arteries. Each plexus is located near the effector organ or region it serves.

In the thorax the cardiac plexus supplies the heart with autonomic input. The pulmonary plexus serves the lungs and bronchial tree.

In the abdomen is the celiac plexus, which innervates the major abdominal organs (stomach, spleen, pancreas, liver, adrenal glands). The two mesenteric plexuses supply the small and large intestines with autonomic nerves. The renal plexus supplies the kidneys and ureters.

Function of the Sympathetic Nervous System

The operation of the sympathetic nervous system is most evident when we are frightened or excited. Sympathetic output supports processes that are required for vigorous physical activity. Therefore, the sympathetic nervous system is called the "fight or flight" system.

"Fight or flight" responses to stress include: an increase in heart rate and blood pressure, an increase in blood flow to skeletal muscles, dilation of bronchial tubes to allows more oxygen intake, break down of glycogen stored in the liver to obtain glucose for energy, and dilation of the pupils. These are actions are all stimulated by the sympathetic nervous system. But the sympathetic nervous system, at the same time, slows down some physiologic functions that are unnecessary drains of

TAKING A CLOSER LOOK

Sympathetic Nervous System

Dilates pupils

Inhibits salivation

Relaxes bronchi

Accelerates heartbeat

T1

Inhibits peristalsis and secretion

Stimulates glucose production and release

T12

Secretion of adrenaline and nonadrenline

Inhibits bladder contraction

Affects the reproductive system

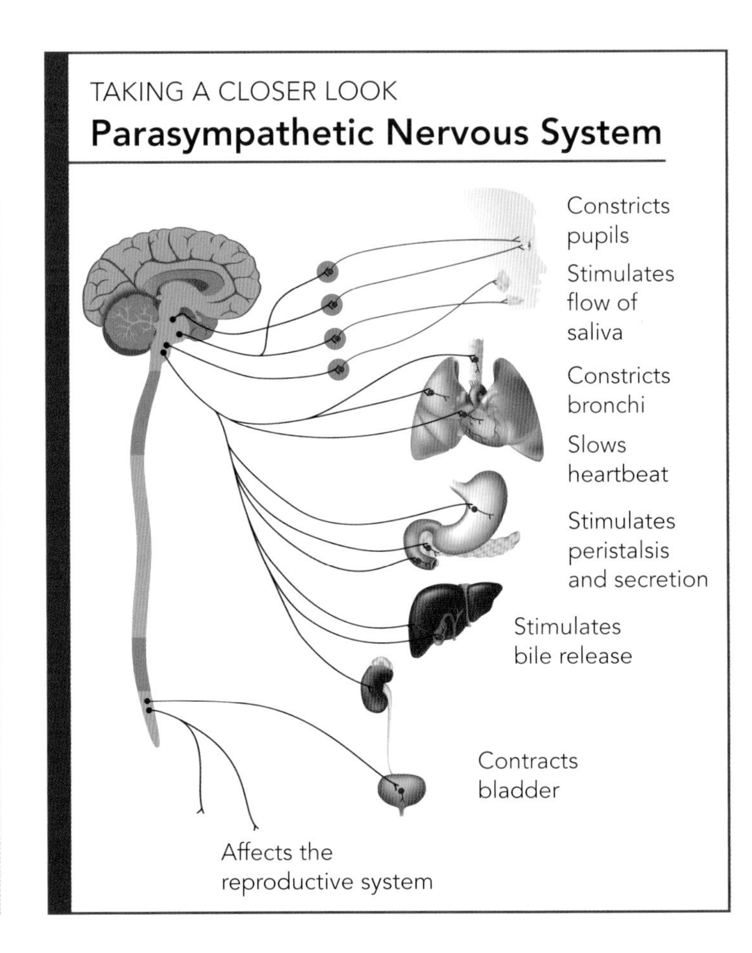

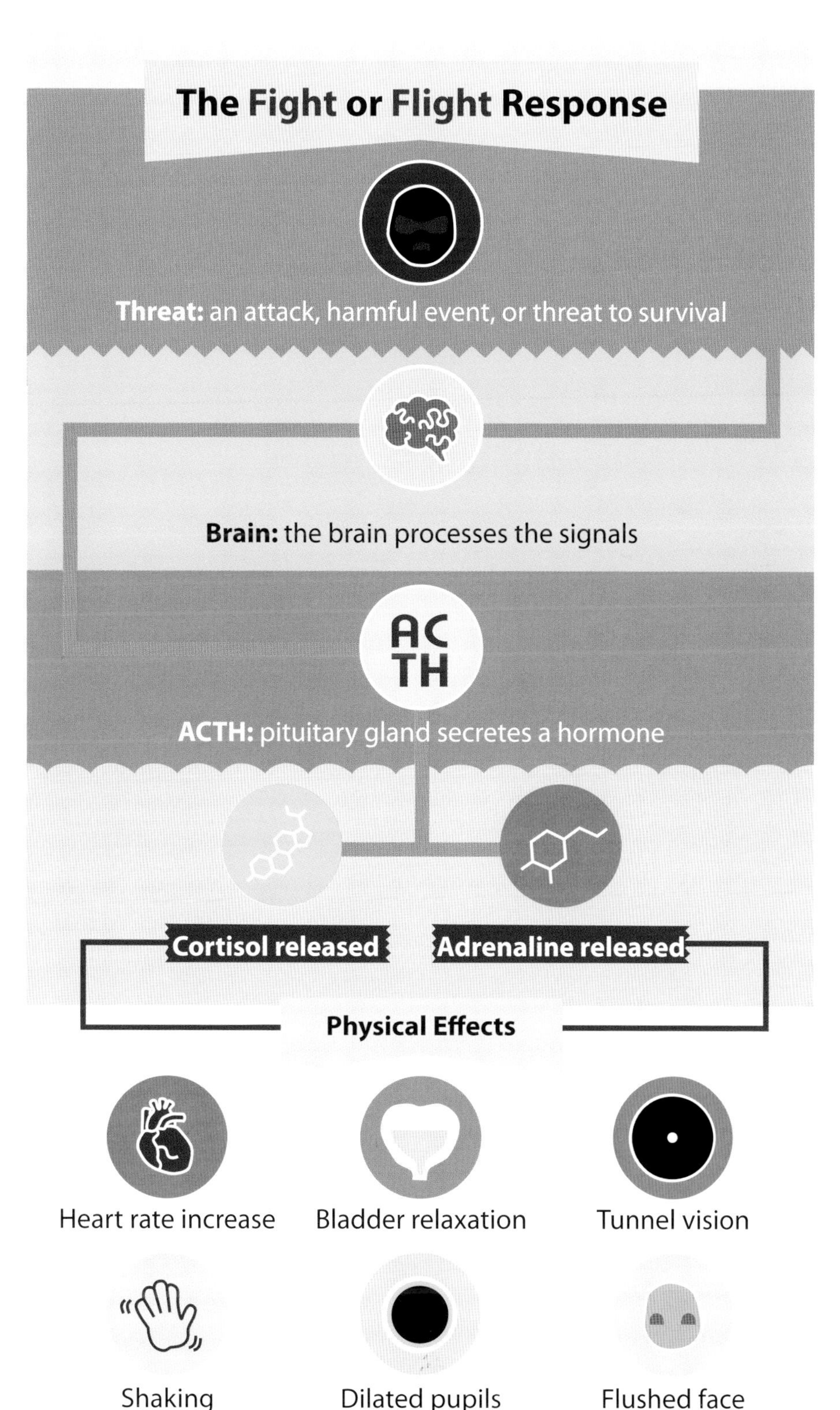

valuable energy when fight or flight is necessary. Thus the sympathetic nervous system can decrease in blood flow to the digestive tract and kidneys, systems not essential to urgent physical activity.

Function of the Parasympathetic Nervous System

In contrast to the sympathetic nervous system, the parasympathetic nervous system could be called the "rest and digest" system. This system is geared to support the rest and recuperation activities of the body. For the parasympathetic nervous system, the support of functions that result in conservation and storage of energy is the goal.

During times of rest, parasympathetic signals to the gastrointestinal tract are often active. These signals promote digestion of food and elimination of waste products. All this aids in preparing the body for any upcoming episode of activity.

SPECIAL SENSES

How many "senses" do we have?

The usual answer is five. So often we hear about using our "five senses." These five senses are taste, smell, touch, sight, and hearing. And it's true. We do smell things. We can taste stuff. We can feel things. We can see objects. We hear sounds.

Some people suggest that we have many more than just five senses. This is true also. It just depends on how you define the concepts of senses.

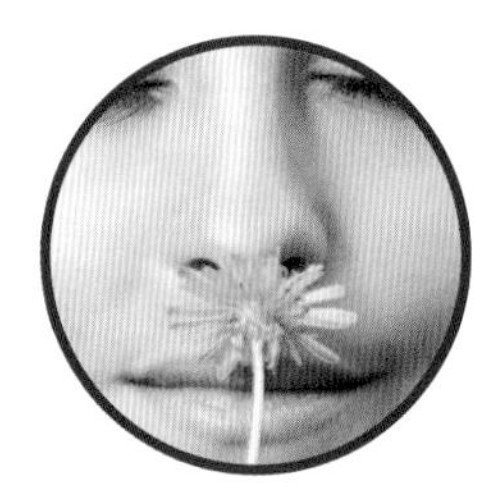  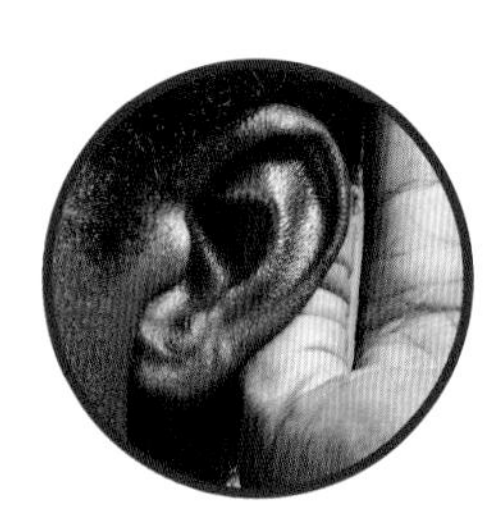 

Remember all the different sensory receptors we listed earlier? That was quite a list! There were probably even a few words—a few "senses"—that you'd never heard of. Mechanoreceptors sense mechanical stress, such as pressure or stretch. Chemoreceptors respond to chemical changes, such as changes in pH (acidity) or the presence of various molecules or ions. Photoreceptors sense light. Thermoreceptors respond to temperature changes. Proprioceptors sense position change. And this list could go on.

Aren't all these things "senses"? Yes, they are. But we don't have to let those senses complicate things by crowding into the classical list of senses.

Let's think instead in terms of what is called a "special sense." These special senses involve sensory receptors contained in specialized organs or structures in the body. Sight, hearing, and taste are examples of special senses.

Specialized photoreceptors are contained in the retina of the eye. Specialized chemoreceptors that allow us to taste are present in the tongue. These are very special types of receptors, and they are confined in organs specifically designed for them. The special senses are taste, sight, smell, hearing, and balance (or equilibrium).

The other sense modalities are better thought of as "general senses." The general senses utilize sensory receptors scattered throughout the body.

But what about touch? Isn't it a sense? Yes, it is, but touch requires the input of many receptors that are not localized in one area or in one organ. Touch is certainly a real sense, but it is better to consider it more of a general sense.

Smell

We are surrounded by smells. Cookies baking in the oven, the smell of freshly cut grass, the pungent aroma of your gym shoes—lots of things have smells. Some things don't, but that's another story. We are going to sniff out the facts about our sense of smell. (See what I did there?)

The sense of smell is called olfaction. It is mediated by special cells, cleverly enough called olfactory sensory neurons. These neurons are embedded in the olfactory epithelium in the roof of the nasal cavity. Air entering the nose passes by the olfactory epithelium on its way to the airways that take it to the lungs.

The olfactory epithelium covers only one-and-a-half square inches but contains millions of olfactory sensory neurons. As you can see in the illustration, the olfactory epithelium rests on the cribriform plate of the ethmoid bone. There are columnar (column-shaped) cells in this layer. These cells act as a support structure. Between the columnar cells are the olfactory sensory neurons. These olfactory neurons are unusual in that they are bipolar neurons. That it, each has only one dendrite and one axon.

The dendrite of an olfactory neuron ends in several cilia. These cilia lie on the surface of the olfactory epithelium. They are covered and protected by a thin layer of mucous.

The bundles of axons of the olfactory neurons extend through holes in the cribriform plate. These bundles of axons make up the olfactory (I) nerve, the first cranial nerve. After passing through the cribriform plate, they synapse with neurons in the olfactory bulb.

At the base of the olfactory epithelium is one more kind of cell. This is called a basal cell. (Clever name, right?) The basal cells of an epithelium are the cells at its base, in the bottom layer. You might want to have a look back at Volume 1 of the *Wonders of the Human Body* for a refresher on epithelium. The basal cells here are actually stem cells that divide to produce replacement olfactory sensory neurons. The olfactory neurons live for only a few weeks. A mechanism for replacing them regularly is vital.

TAKING A CLOSER LOOK

Gross Anatomy of the Nasal Cavity

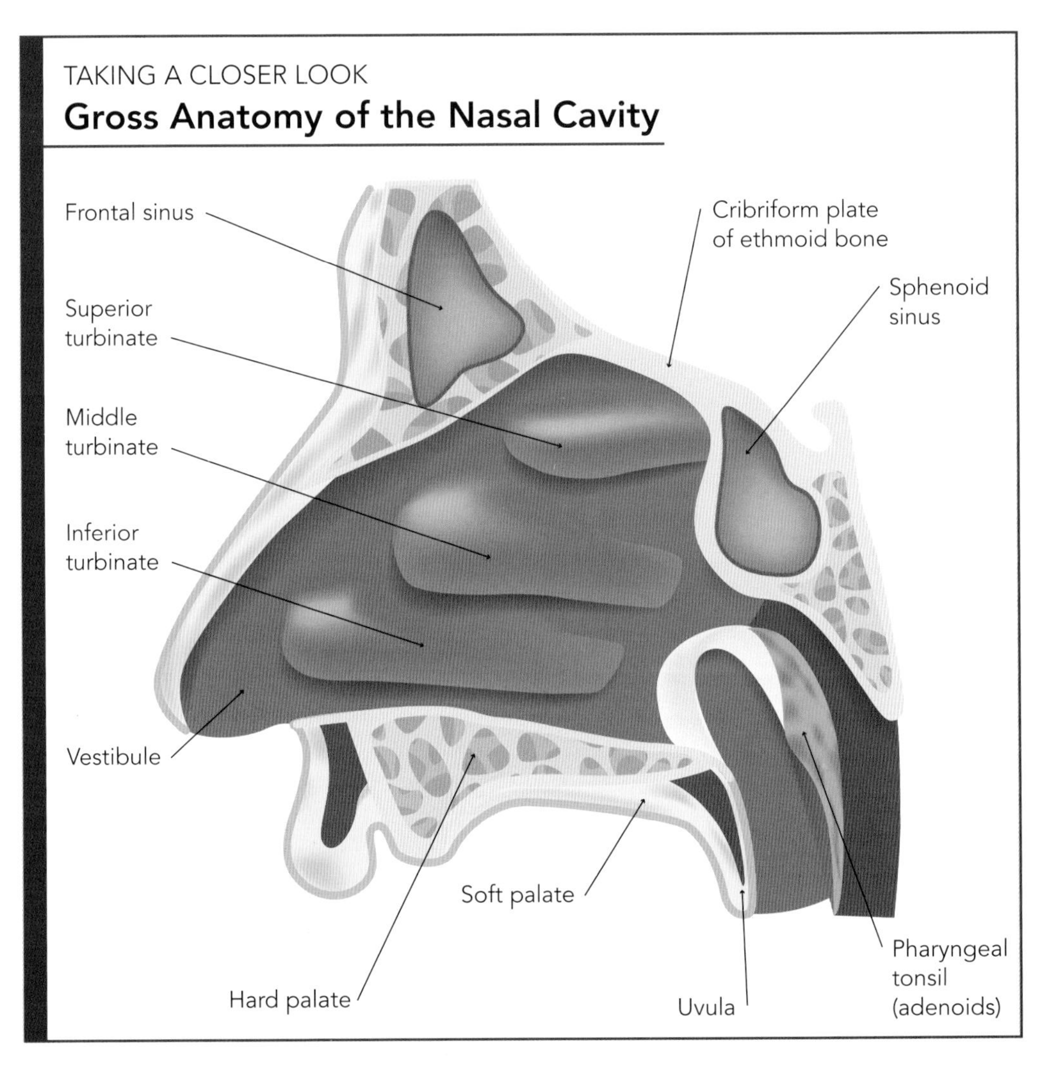

How Does Smelling Work?

Now that we know the anatomy, how does smelling work?

Substances that can trigger smell are called odorants. In order for an olfactory sensory neuron to be triggered, an odorant molecule must reach its receptor. This occurs after the odorant molecule is breathed in. As this molecule passes the olfactory epithelium, it is trapped by the mucous layer covering the epithelium.

In the mucous layer the odorant contacts the olfactory cilia and binds to a receptor site. This binding triggers the opening of ion channels nearby. If enough odorant molecules trigger the receptor cell, an impulse is sent down the entire length of the neuron.

Once a full action potential is triggered, the nerve impulse reaches the olfactory bulb. Then the signal moves down the olfactory tract to the olfactory area of the cerebral cortex. We then perceive a smell!

We do know that our threshold for smell is low. That means it only takes a few molecules of some odorants to trigger smell. Certain odors we can perceive even at extremely low concentrations (like 1 molecule in 50 billion!) while other odorants require a higher concentration for perception.

For many years the traditional thinking was that humans could detect about 10,000 different odors. A recent study has challenged that number. This new study claims that humans may be able to detect as many as one trillion smells (again, who counted them?)! Research is obviously ongoing.

And walk in love, as Christ also has loved us
and given Himself for us,
an offering and a sacrifice to God
for a sweet-smelling aroma.

(Ephesians 5:2)

TAKING A CLOSER LOOK

How Smelling Works

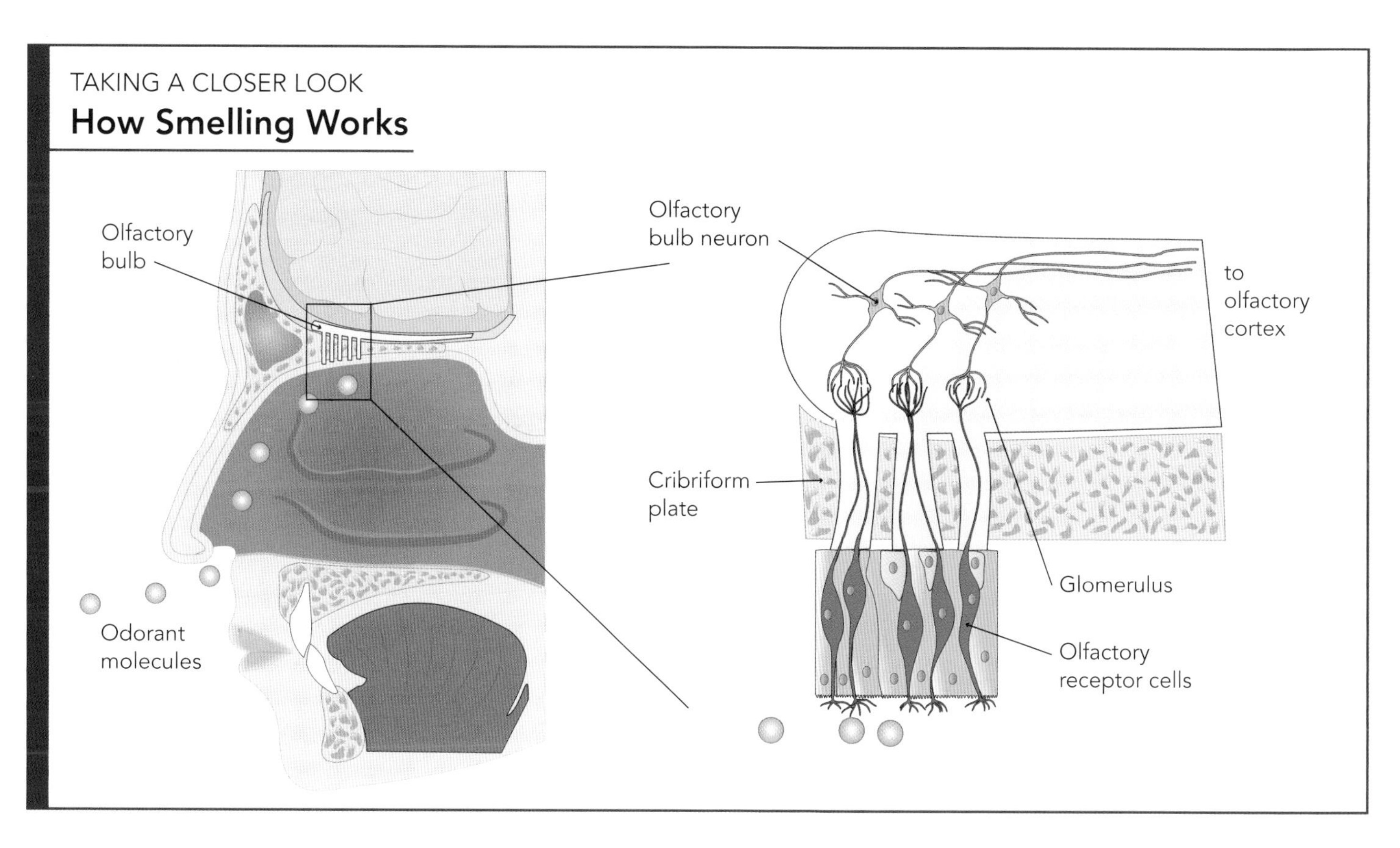

Taste

And the house of Israel called its name Manna.
And it was like white coriander seed,
and the taste of it was like wafers made with honey.
(Exodus 16:31)

The sense of taste, like the sense of smell, is a chemical sense. Our ability to taste depends on our ability to detect, and then react to, certain chemicals in our environment. We can do this by means of our taste buds.

If you look at your tongue in the mirror, you will see that it has a rather rough-looking surface. This appearance is due to the presence of many protuberances called papillae on its surface. Filiform (thread-like) papillae cover the major portion of the front two-thirds of the tongue. Unlike the tongue's other papillae, these lack taste buds. Fungiform papillae are mushroom-shaped and are scattered over the entire surface of the tongue. Taste buds are found on the top of fungiform papillae. Foliate papillae are leaf-like folded ridges on the sides of the tongue near the rear. Nodular-appearing circumvallate papillae form a row across the back of the tongue. In these two types of papillae, Taste buds are found in the side walls of the foliate and circumvallate papillae.

In the olfactory epithelium, the dendrite of the olfactory neuron was the sensory receptor, but taste buds are different. In a taste bud, the neuron is not the sensory receptor. The actual sensory cell is the gustatory epithelial cell. (Gustatory means "taste." Bet you guessed that.) Each of these cells has microvilli that extend through a taste pore on the epithelial surface. These microvilli are the sites that trigger a reaction in the gustatory cells. Also in contact with the gustatory epithelial cell are sensory neurons. These are the neurons that start a nerve signal on its way to the brain.

Because of their location, gustatory epithelial cells are subjected to lots of wear and tear. They are thus easily damaged and have a very short life span, usually just a week or so. Just as the basal cells in the olfactory epithelium divide to make replacement cells, so the basal epithelial cells in the taste buds divide and produce replacement gustatory epithelial cells regularly.

Physiology of Taste

As mentioned, the sense of taste is a chemical sense. When exposed to a certain chemical, a specialized receptor reacts to the stimulus and ultimately a nerve signal is produced. Even though smell and taste are both chemical senses, the process works differently for taste.

Remember the actual receptor cell for taste is the gustatory epithelial cell. This cell is not a nerve cell.

TAKING A CLOSER LOOK

Tongue

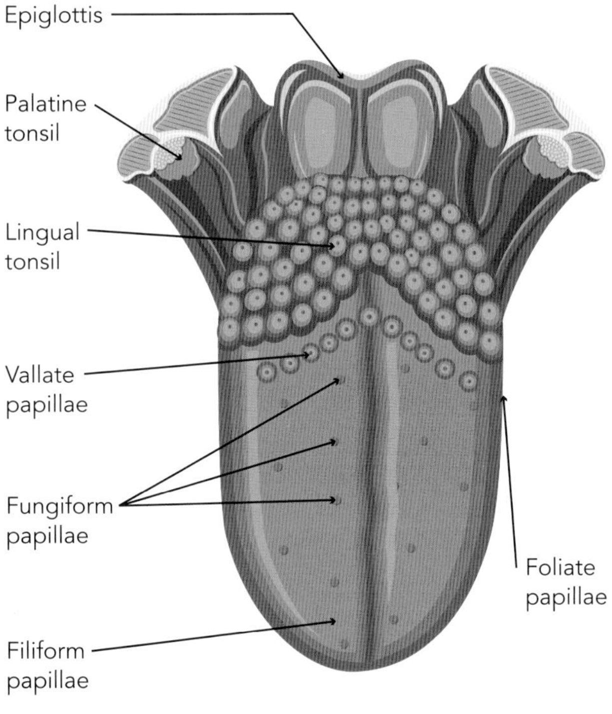

TAKING A CLOSER LOOK
Taste Buds

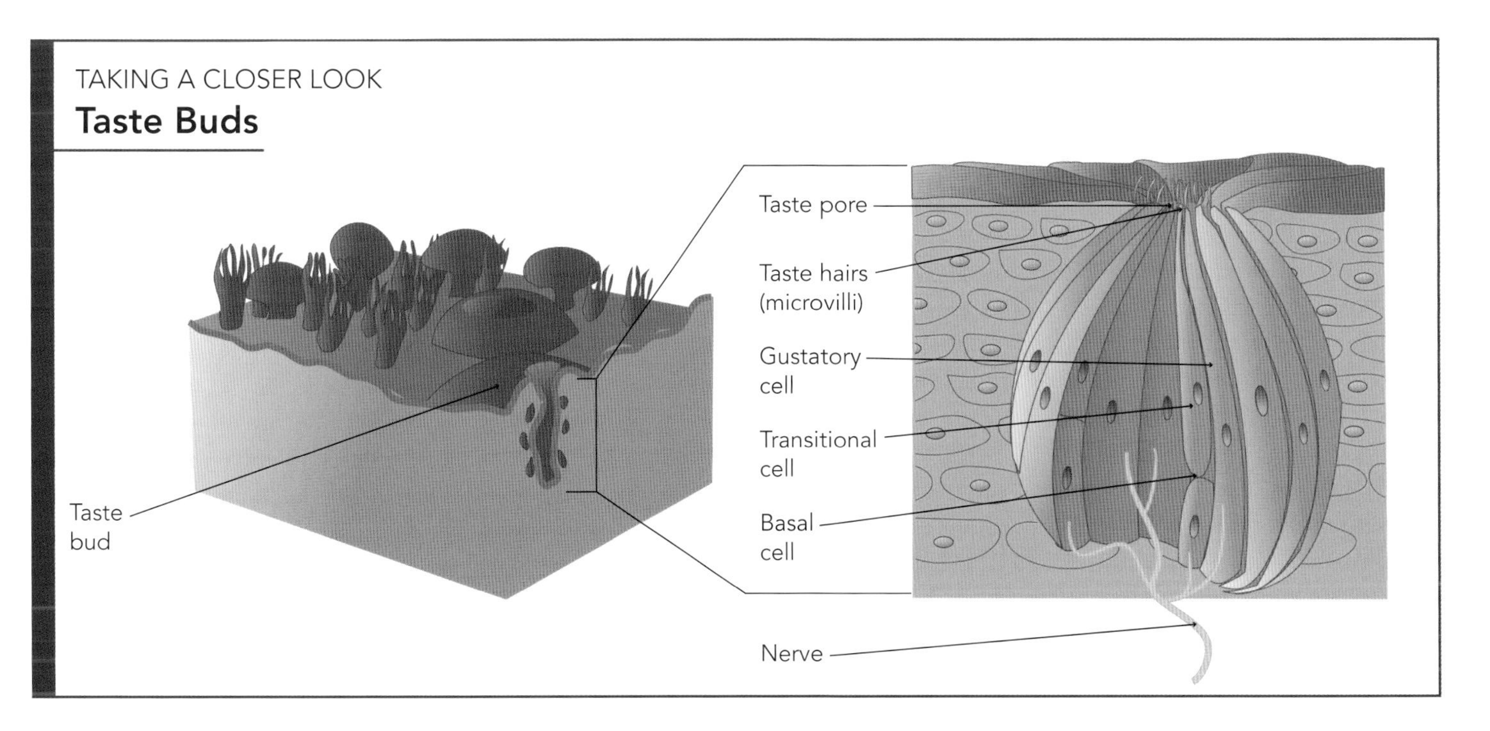

That's a big difference right there. So what gives? Well, it's really easy.

A gustatory epithelial cell is exposed to a stimulatory chemical called a tastant. This happens at the microvillus at the taste pore. When the tastant binds to the chemical receptor on the microvillus, no nerve signal is generated. The gustatory cell is not a nerve cell. What happens is that the gustatory epithelial cell releases a neurotransmitter that then stimulates receptors in the dendrites of the sensory neurons that are in the taste bud. This then triggers the action potential in the neurons.

The afferent (sensory) fibers that carry taste signals to the brain are mainly in two of the cranial nerves. The facial nerve (VII) carries impulses from the anterior two-thirds of the tongue, and the glossopharyngeal nerve (IX) carries impulses for the rear third of the tongue. Fibers from these nerves synapse in the solitary nucleus of the medulla oblongata. Then the signals are taken to the thalamus and ultimately on to the gustatory cortex.

Types of Tastes

At present, taste sensations have been categorized into five basic groups:

1. Bitter taste is perceived as unpleasant. Bitter taste can result from bases or alkaloids (like the medication quinine, for example). This is the most sensitive of the taste modalities.

2. Sweet taste is generally pleasant. Sugars are an obvious source for this taste. Other sources include some alcohols and some amino acids.

3. Sour taste detects acidity. Citrus fruits tends to be naturally acidic.

4. Salty taste detects inorganic salts. Sodium chloride (table salt) is the best example.

5. Umami is the taste sensation produced by the amino acids, glutamate and aspartate. Some describe this as the "savory" taste. Beef and cheese can elicit this taste.

There is one other "taste" that may be added to the list in the future. This is, at present, called oleogustus. This is the taste for fats. So far this has not been universally accepted, but over time it may be the "sixth" taste.

Every part of the tongue can detect any of the taste modalities. Sweet sensation or sour sensation are not localized to specific area of the tongue as has been previously taught. Any taste bud can detect any of the different tastes, However, each individual gustatory epithelial cell apparently can only be triggered by a single taste modality. Fortunately, there are many different gustatory cells in every taste bud!

Another factor greatly influencing how we perceive a taste is our sense of smell! If our olfactory sense is triggered when we eat, it enhances the pleasure our food gives us. On the other hand, if you have ever had a really bad cold or an allergy attack, you understand that the opposite of this is also true. When your sinuses are congested, you cannot smell things. Funny how during those times, food just doesn't taste as good!

Hearing

The ear is the organ associated with hearing. It surprises many people that the ear is also the organ responsible for balance (or equilibrium).

These senses are so amazing. We can hear the booming fireworks on the Fourth of July and yet are

TAKING A CLOSER LOOK

Gross Anatomy of the Ear

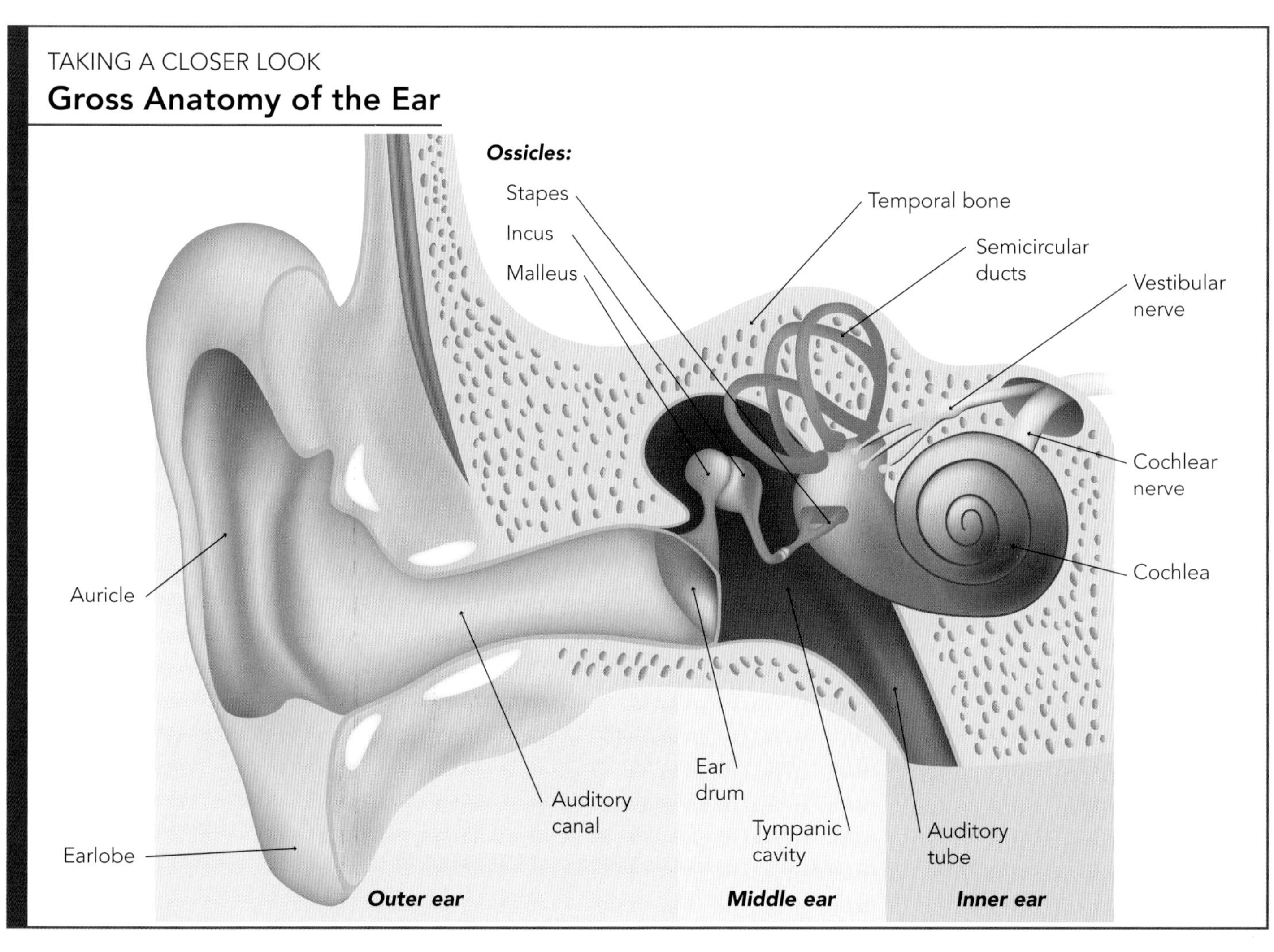

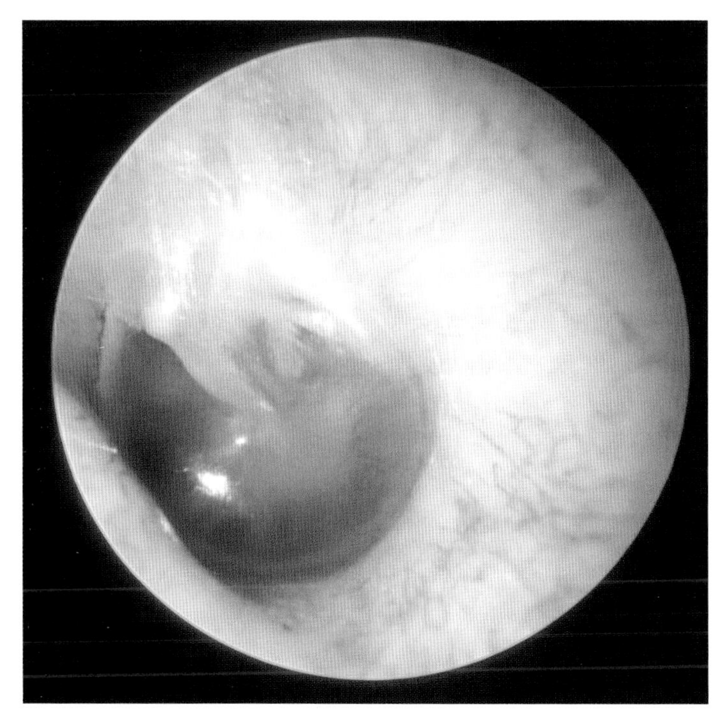
The eardrum

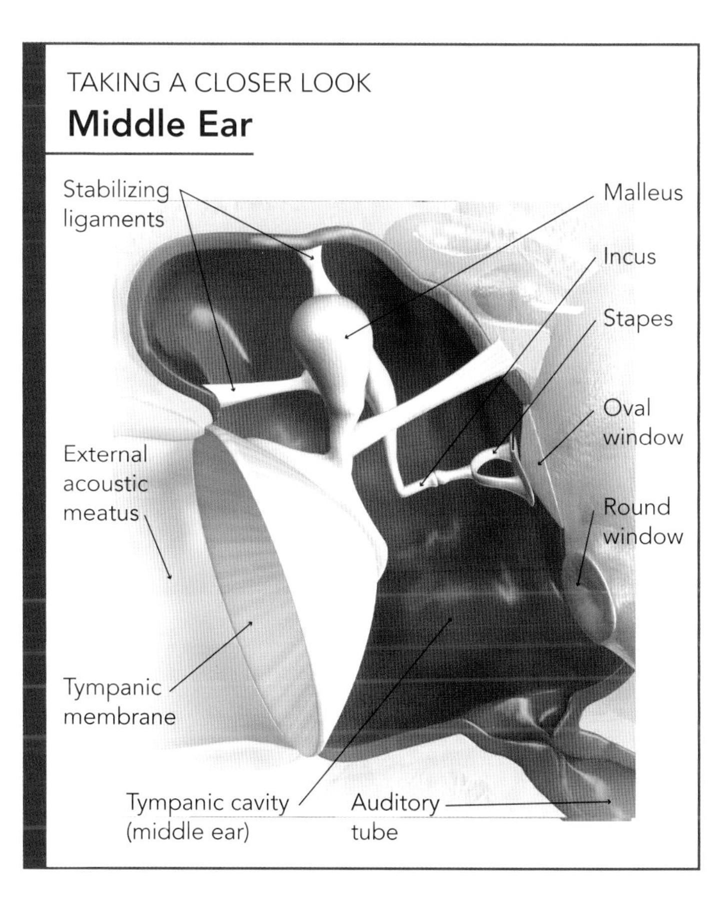

able to heard the gentle buzzing of the bee in our backyard. We can remain standing, balanced, with our eyes closed. We can run the bases and climb the stairs without giving it a thought. Let's see how they work.

Anatomy of the Ear

The ear is composed of three main regions: the external ear, the middle ear, and the inner ear.

The external ear is what most people refer to when they talk about the ear. The external ear is the part that is visible to the world. The shell-shaped protrusion from the side of your head is called the auricle. It is made of elastic cartilage covered by skin. The external auditory canal is a tube through which sound waves move toward the tympanic membrane (eardrum). This auditory canal is about an inch long.

The external auditory canal is lined with skin, a few hairs, and special glands called ceruminous glands. These glands produce cerumen, or earwax. Although earwax can be a problem if it builds up and blocks the ear canal, it is there to keep foreign objects from reaching the delicate tympanic membrane.

At the end of the external auditory canal is the tympanic membrane, or eardrum. It is round and coned slightly inward. As we will see soon, the tympanic membrane vibrates when sound strikes it. The tympanic membrane marks the boundary between the external ear and the middle ear.

The middle ear is a small cavity in the temporal bone of the skull. The tympanic membrane separates it from the external ear. The middle ear is separated from the inner ear by a bony wall containing two openings, the oval window near the top, and the round window below. Both of these windows are covered by membranes. Also opening into the middle ear is the Eustachian tube. This tube connects to the nasopharynx (high in the back of the throat). Each

Size of the stapes.

end of the Eustachian tube is open to air, the one end to the air in your throat, which is pretty much at the same pressure as the air outside your ears in the room, and the other end to the air in the middle ear. Therefore, it helps equalize the pressure on both sides of the tympanic membrane. This equal pressure allows the eardrum to vibrate freely as sound waves strike it.

Strung across the middle ear is a chain of three small bones. In fact, these are the three smallest bones in the body. These bones are named based on their appearance. The first is the malleus (Latin for "hammer"). On one side this hammer attaches to the inner surface of the tympanic membrane, and on the other side it connects to the middle bone in the series, the incus. The incus (Latin for "anvil") then connects to the third bone, the stapes. The base of the stapes (Latin for "stirrup") then fits into the oval window. So you see, the vibrations that start on the eardrum are going to be transmitted to the hammer, then to the anvil, then to the stirrup, and on to the oval window.

The inner ear has two parts, one inside the other. The outermost is the bony labyrinth, which is a series of cavities in the temporal bone. Within this is the membranous labyrinth. The membranous labyrinth is a series of tubes and sacs that lies within the bony labyrinth. Both labyrinths are filled with fluid. Inside the bony labyrinth, surrounding the membranous labyrinth, is perilymph. The fluid inside the membranous labyrinth is called endolymph.

There are three major parts of the bony labyrinth. They are the vestibule, the semicircular canals, and the cochlea.

The vestibule is the bony labyrinth's central chamber. It is located medial to the middle ear. It is separated from the middle ear by the oval window, to which the stapes attaches. The membranous labyrinth

TAKING A CLOSER LOOK

Inner Ear

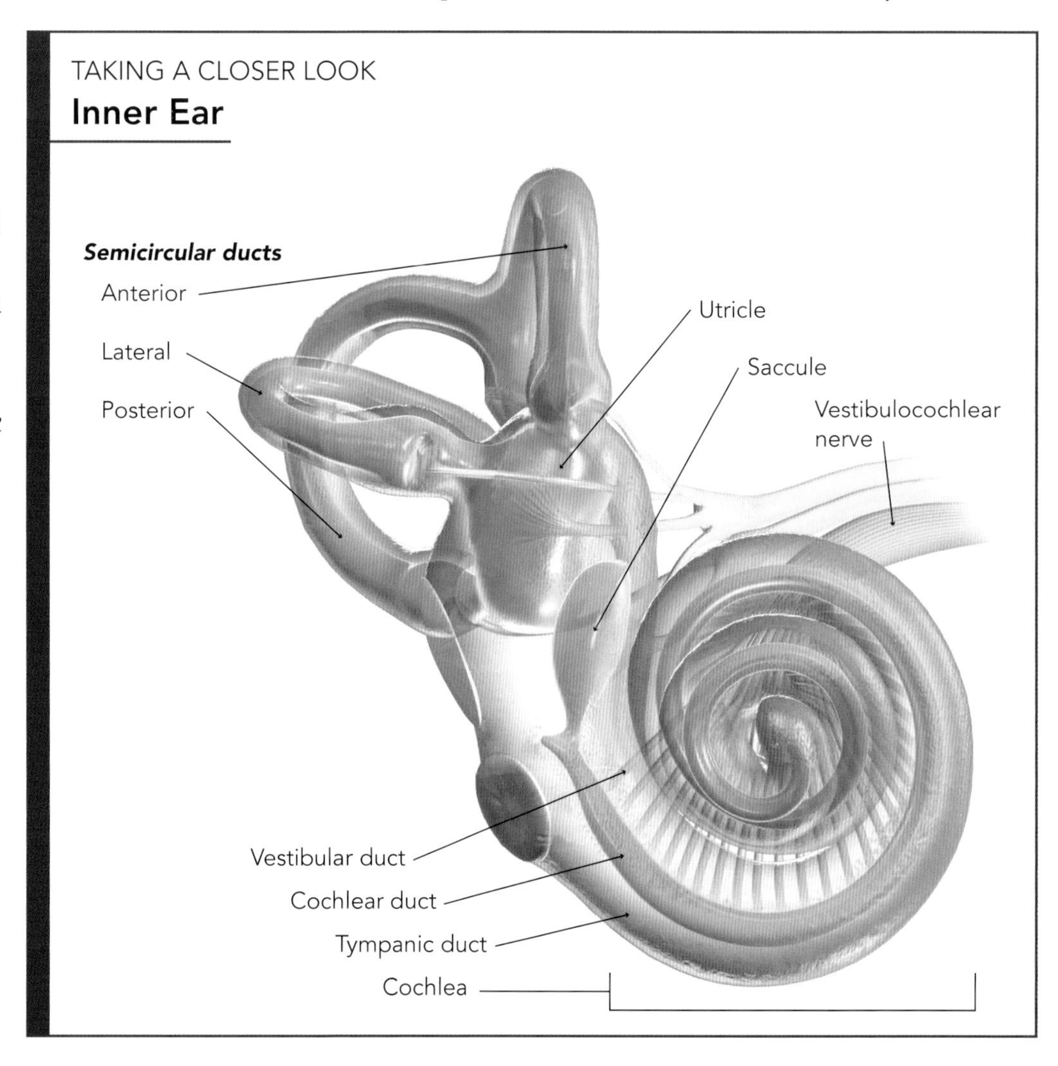

components in the vestibule consist of two sacs, the utricle and the saccule. Both of these are important in controlling equilibrium, as we will see later.

Posterior to the vestibule are the semicircular canals. There are three semicircular bony tubes oriented are right angles to one another. Through each of these canals runs a duct called a semicircular duct. At the end of each duct is an enlarged area called the ampulla. These ducts—parts of the membranous labyrinth—open into the utricle.

Anterior to the vestibule is the cochlea. This is a spiral chamber made of bone. It is shaped pretty much like a snail's shell. Running through the middle of the cochlea is the cochlear duct. Inside the cochlear duct is the spiral organ (also called the organ of Corti) which is the hearing receptor. By its placement in the cochlea, the cochlear duct effectively divides the cochlea into three chambers. These chambers are called scala. There is the scala vestibuli, which begins at the oval window. The scala media is the cochlear duct itself. Finally, there is the scala tympani, which ends at the round window.

At the distal portion of the scala vestibule is a small opening, called the helicotrema. This small opening connects the fluid in the scala vestibule and the scala tympani.

A closer look at a cross-section of the cochlea shows these three chambers. The cochlear duct is separated from the scala vestibuli by the vestibular membrane. The floor of the cochlear duct, separating it from the scala tympani, is the basilar membrane.

Resting on the basilar membrane is the spiral organ (or organ of Corti). This organ is composed of supporting cells and cochlear hair cells. The hair cells are arranged in rows and are covered by the tectorial membrane. At the base of the hair cells are fibers of the cochlear branch of the vestibulocochlear (VIII) nerve.

Yep, that's an awful lot of anatomy, but we will make sense of all this very soon. Hang in there, and keep a close eye on the illustrations as we go through the path of sound and the way it gets transmitted to your brain.

Sound

When we hear something, we are sensing sound waves from the environment. Understanding what is happening when we hear requires some knowledge about sound itself.

Sound is a series of vibrations. It can be illustrated as a series of waves as shown. These vibrations are pressure waves, and they must travel through a medium, such as air or water. Because they can vibrate, even the bones in the skull are capable of transmitting sounds waves. You can compare sound waves to the ripples you produce when you throw a rock into a quiet pond. Sound cannot travel through a vacuum, such as in space.

Different sounds sound different, right? Otherwise, there would be only one sound. Sounds can vary in pitch or in amplitude. Pitch is the frequency of the sound. The more waves per second, the higher the pitch. The sound from a bass drum has a much lower frequency than the sound from a policeman's whistle. The whistle has a much higher frequency.

Sounds also different in their loudness, or intensity. A bass drum can make a soft sound or a very loud

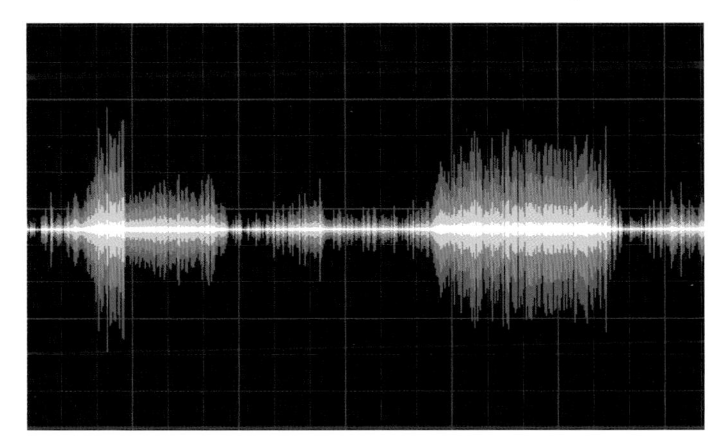

one. The frequency of the two sounds will be very close to the same, but the amplitude of the sounds will differ. In the illustration, the greater the size of the wave, the greater the amplitude. The greater the amplitude, the louder the sound. Take a train whistle as an example. If a train is one half mile away and blows its whistle, you will hear it and most likely recognize it as a train whistle. It will be distinct, but not comfortably loud. If that same train blows its whistle when it is only fifty yards away, the sound will be very loud. When it is louder, it has a higher amplitude.

Both of these concepts, pitch and amplitude, are important as we explore our sense of hearing.

> *"I have heard of You by the hearing of the ear,*
> *But now my eye sees You.*
> (Job 42:5)

TAKING A CLOSER LOOK

Auditory Path

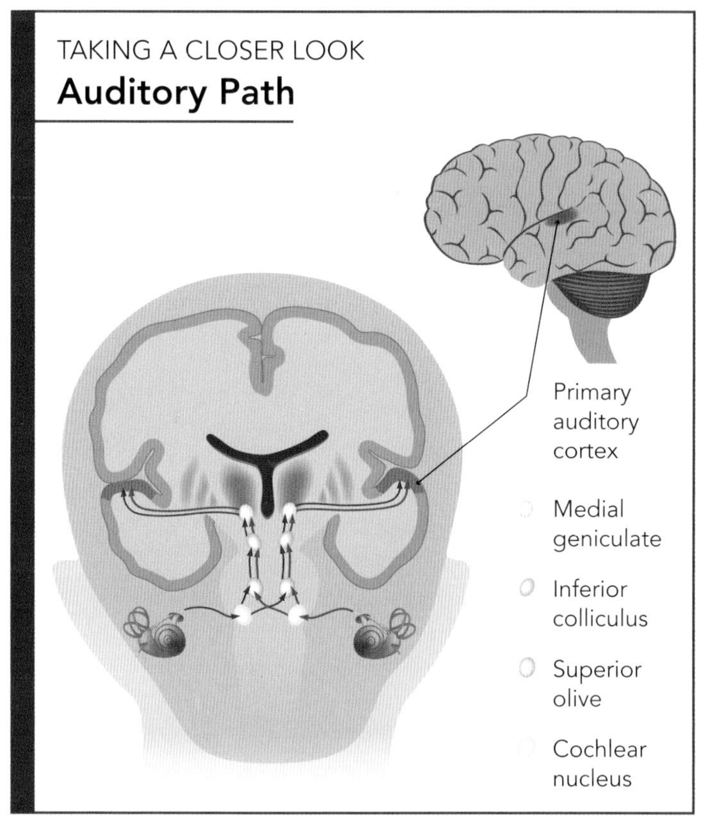

TAKING A CLOSER LOOK

How Hearing Works

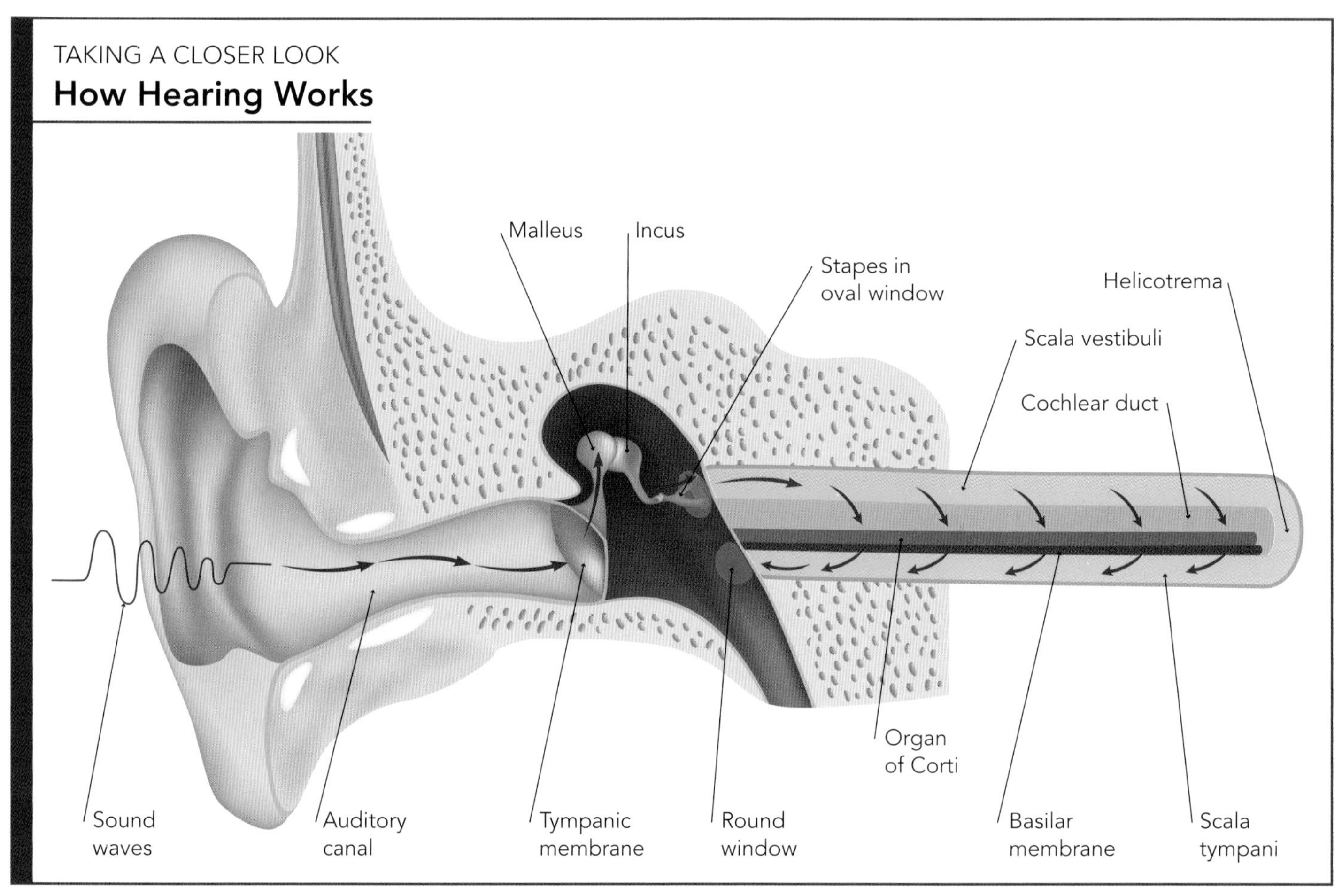

How We Hear

Hearing is our ability to convert the pressure waves (sound waves) in our environment to action potentials that can be transmitted to the brain. In the brain, these nerve signals are processed and perceived as sound.

We will follow a sound wave, step by step, though the ear.

1. The external ear (the auricle and external auditory canal) captures sound waves (pressure waves) and directs them toward the tympanic membrane.

2. When it reaches the tympanic membrane, a sound wave strikes the membrane and causes it to vibrate at the same frequency as the sound wave. This matching of the frequency transmits the pitch of the sound. More than that, the amplitude of the sound wave is transmitted also. The louder the sound reaching the tympanic membrane, the farther the membrane is pushed as it vibrates. So it is not only how fast the membrane vibrates, but it is also important how far the membrane moves with each vibration, that passes the pitch and amplitude of the sound wave accurately to the middle ear.

TAKING A CLOSER LOOK

Organ of Corti

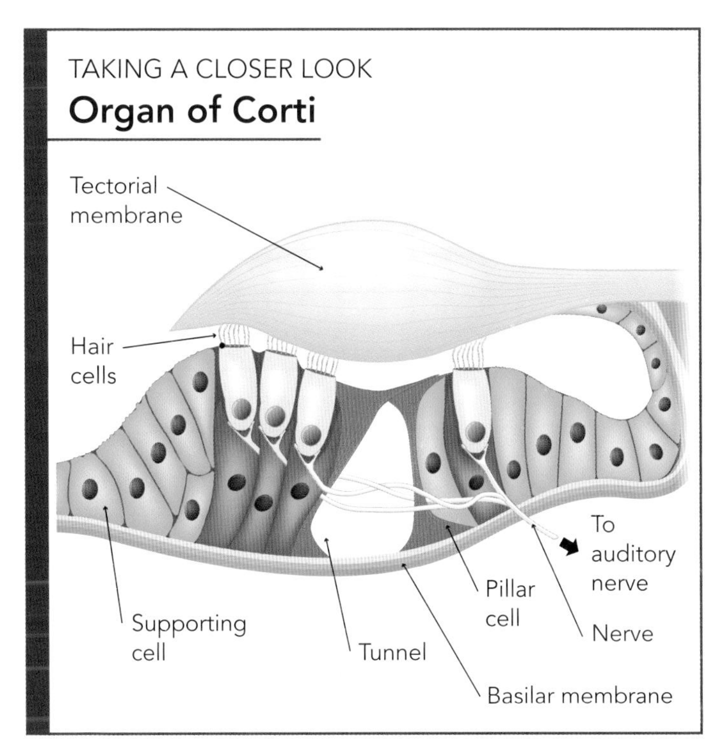

TAKING A CLOSER LOOK

Anatomy of the Cochlea

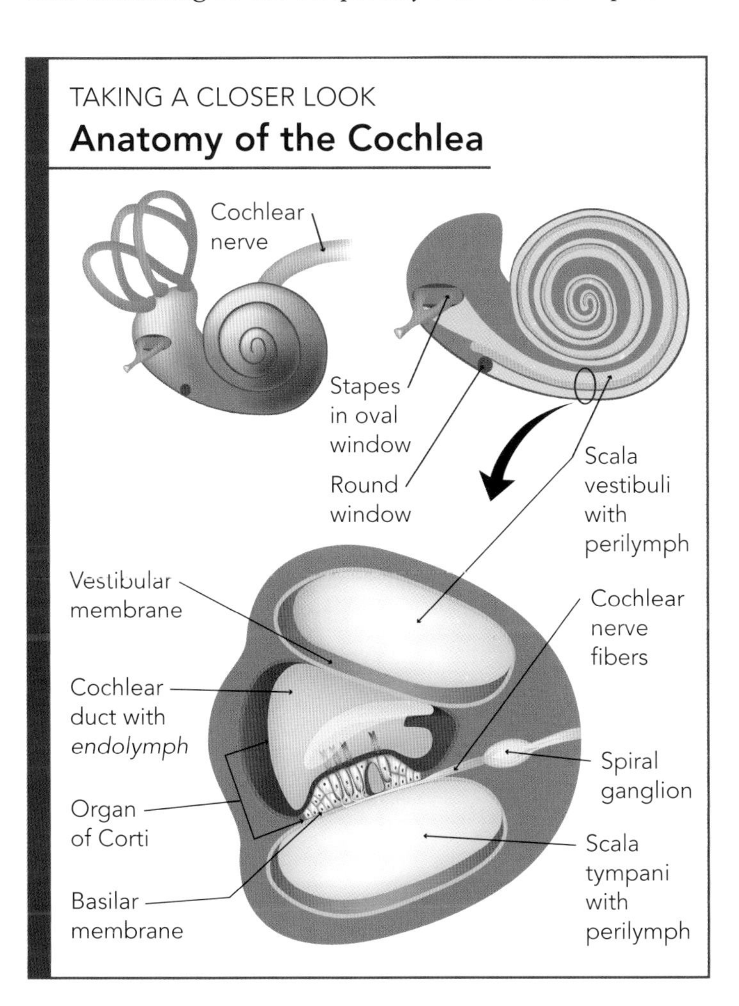

3. When the tympanic membrane vibrates, it causes movement in the three bones of the middle ear—first the malleus, then the incus, and finally the stapes. Both the frequency of the vibration and its intensity are transmitted, via these bones, to the oval window.

4. The base of the stapes rests against the oval window. The vibrations transmitted from the tympanic membrane are thus transferred to the oval window, and it begins to vibrate accordingly. These vibrations are now transmitted onward, but now the transmission is through fluid, not air (as in the external ear) or bones (as in the middle ear). The pressure waves from the oval window are now carried by the perilymph. These waves travel down the scala vestibuli , through the helicotrema, and back down

the scala tympani. When the waves reach the end of the scala tympani, they reach the round window. The pressure waves cause the round windows to flex and bulge. This bulging effectively damps out and ends the pressure waves.

5. As the perilymph carries the sound wave throughout the scala, it causes the basilar membrane to vibrate. (Keep your eye on those illustrations as you trace the path of sound's vibrations!) When the basilar membrane vibrates, the cochlear hair cells move against the tectorial membrane. The movement of the hair cells generates receptor potentials.

6. The receptor potentials produced by the cochlear hair cells trigger action potentials in the neurons associated with them. These nerve impulses are then carried by the cochlear branch of the vestibulocochlear (VIII) nerve.

Another fascinating aspect of the sound transmission process is how the ear is able to process different frequencies efficiently. It has to do with the structure of the basilar membrane. The basilar membrane is "tuned" to respond to different frequencies along its length. On the end nearer the oval window, the membrane responds better to higher frequencies. On its more distal end, the membrane responds better to lower frequencies.

Balance

It is incredible that we don't fall down all the time. It takes an enormous amount of information processing to keep all the right muscles contracting at just the right time with just the right amount of force...to keep us from falling down. All this cannot happen if we don't know where we are and what

TAKING A CLOSER LOOK

Vestibular System

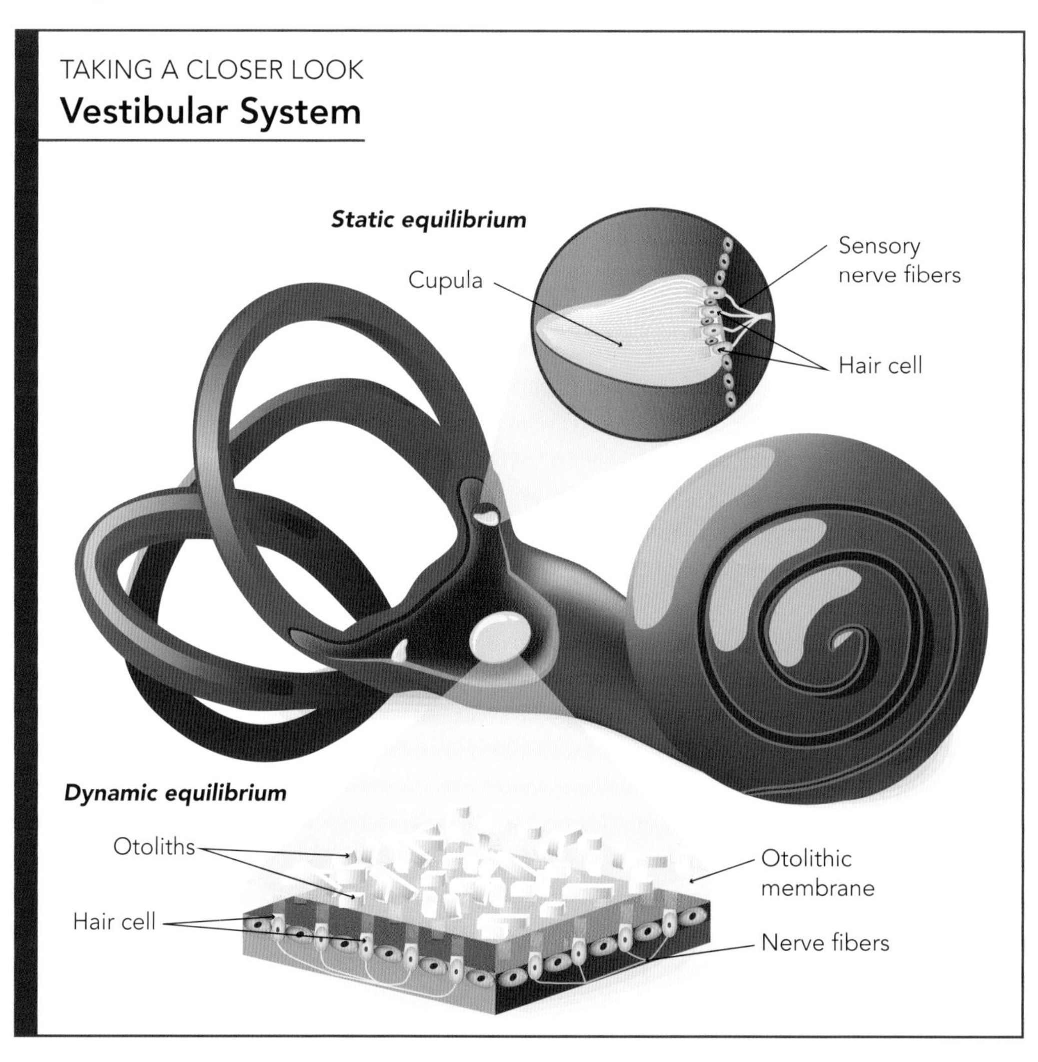

direction we are moving in right now. Is our head up or down? Are we looking to the left or to the right? There must be something that tells the brain all this. That something is called the vestibular apparatus.

The vestibular apparatus consists of the utricle, the saccule, and the semicircular canals. From these structures signals are sent to the brain to keep it informed of the head's position in space.

In the walls of both the utricle and the saccule is found a structure called a macula. The utricle and the saccule are positioned perpendicular to each other (at a 90-degree angle to one another). Therefore they are able to monitor movement in two different planes.

A macula is a layer of two different types of cells: supporting cells and hair cells. The supporting cells do exactly what their name suggests; they provide support. The hair cells have microvilli and are the sensory receptor cells. The hair cells are covered by a jelly-like layer called the otolithic membrane. In the otolithic membrane are many small stones (actually calcium carbonate crystals) called otoliths. The otoliths add mass to the otolithic membrane. When the head moves, the otolithic membrane, weighted down by these little rocks in your head, is pulled by gravity. As the membrane moves, the hair cells underneath are moved. This movement triggers impulses in the nerve cells near the base of the hair cells. These are fibers of the vestibular portion of the vestibulocochlear (VIII) nerve.

But there's more! Balance is pretty important. These semicircular canals—another vital part of the vestibular apparatus—are also very important in maintaining balance.

If you recall, the three semicircular canals are oriented at right angles to one another. Each canal opens into the utricle. In the ampulla of each canal is found a receptor called the crista ampullaris (or crista, for short). The structure of the crista is very similar to the structure of the macula we just studied. There are supporting cells and hair cells. Again, the support cells support the hair cells, and the hair cells are the sensory cells. Covering the crista is another jelly-like mass. This is called the cupola.

As the head rotates, the endolymph in the semicircular canals moves also. As the endolymph moves, it moves the cupola, which in turn moves the hair cells. When the hair cells move, receptor potentials are generated, which then trigger action potentials in their associated neurons. Again, the nerve impulses are carried by the vestibular branch of the vestibulocochlear (VIII) nerve.

Quite an elaborate system for keeping our balance, wouldn't you say?

Vertigo

Vertigo is the sensation a person experiences when he feels like he is moving but he isn't. Vertigo can be spinning sensation or a falling sensation. Very often this dizziness is accompanied by nausea, breaking out in a sweat, or feeling faint. Movement of the head to either side usually makes the symptoms worse.

The most common disease that results in vertigo is benign paroxysmal positional vertigo. This is probably due to loose otoliths moving into a semicircular canal. These loose otoliths are felt to cause abnormal movement of the endolymph, triggering the vertigo.

Another cause of vertigo is labyrinthitis. Labyrinthitis in an inflammation of the inner ear. It may be caused by a virus.

Other possible causes of vertigo include Ménière's disease, migraine, stroke, multiple sclerosis, and Parkinsonism.

Sight

I don't have to tell you where you would be without your eyes. In the dark, that's where. We use our eyes for things we take for granted every day. Reading a book, watching a movie, seeing your mom's smiling face, seeing your dad's frowning face when you forgot to do your homework (well, you can probably do without that one...)—without eyes you could not really picture what you'd be missing. From the time you wake up in the morning until the time you fall asleep at night, your eyes are involved in almost everything you do.

There is an enormous amount of visual information that we process each day. This is how it all works.

Now therefore, stand and see this great thing which the Lord will do before your eyes.

(1 Samuel 12:16)

Corneal Abrasions

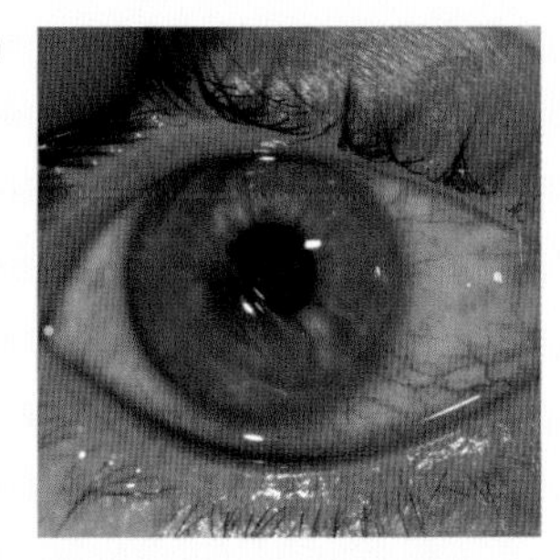

A corneal abrasion is a scratch on the surface of the cornea. As the cornea has many nerve endings, a corneal abrasion can be quite uncomfortable, causing pain, redness, and sensitivity to light.

Corneal abrasions are very common and are usually caused by a finger poked in the eye. Abrasions are also caused by windblown objects or even contact lenses.

There is no specific treatment, although some people have some degree of pain relief if the eye is held shut with a patch. Most patients recover fully in 3-4 days.

TAKING A CLOSER LOOK

Gross Anatomy of the Human Eye

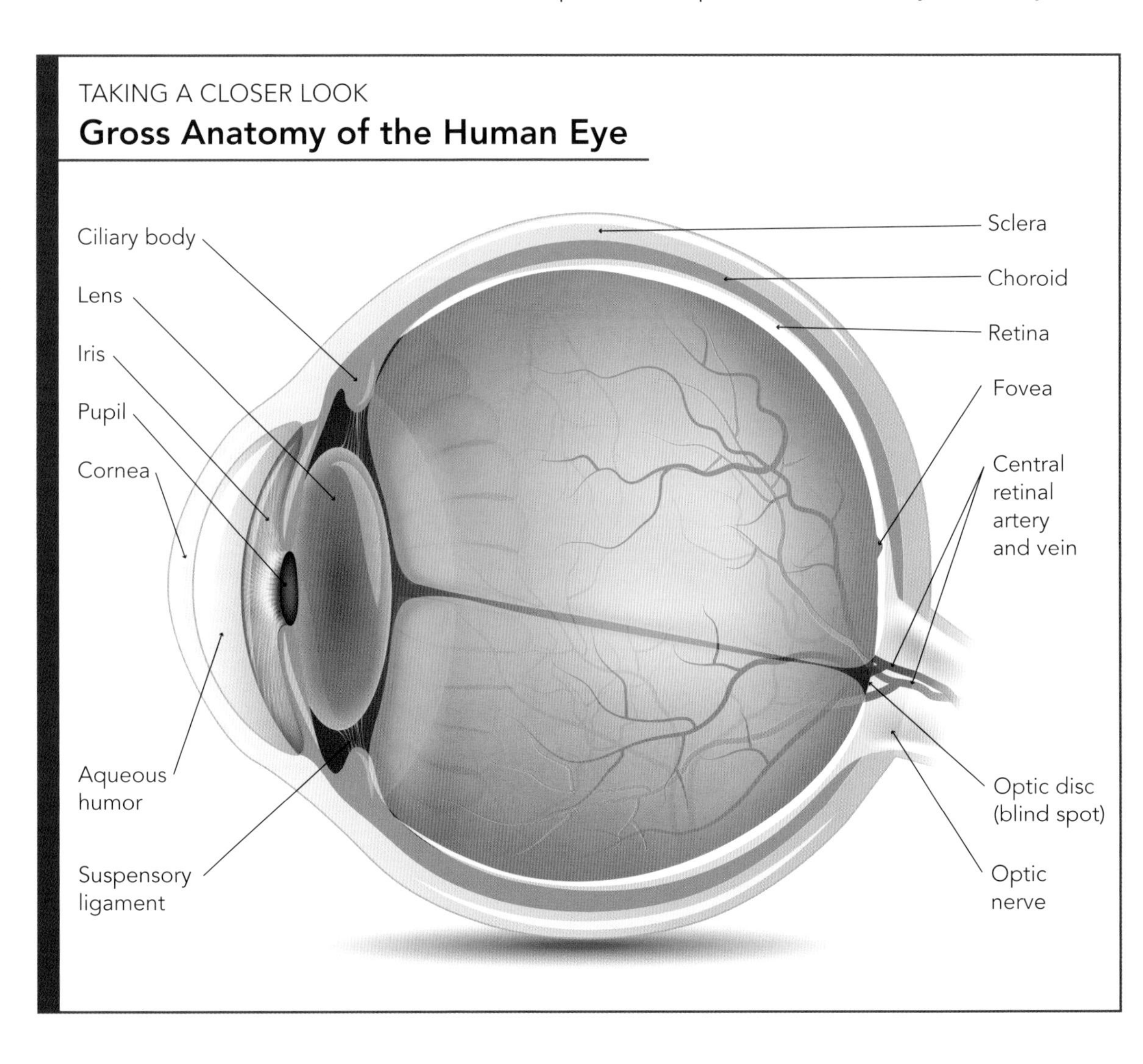

Anatomy of the Eye

The eye is a fascinating thing. It is small compared to other some organs (usually about 1 inch in diameter), but it is quite complex in both its structure and its function.

The outside layer of the eye consists of two parts. The sclera is a layer of dense connective tissue. It is quite durable, and it helps the eye maintain its round shape. When you see the "white" of someone's eye, you are seeing the sclera. The other part of the outer layer of the eye is the cornea. This is the clear portion of the eye in the front. It is a good thing that this portion is clear. This is where the light enters the eye! The cornea is curved in the front. This curvature bends light as it enters the eye, beginning the process of focusing it.

The cornea has lots of nerve endings, so it is very sensitive to being touched or scratched. On the other hand, it has no blood vessels, so it is completely clear.

The middle layer of the eye is called the vascular tunic. The posterior 5/6 of this layer is the choroid. The choroid contains many blood vessels and provides oxygen and nutrients to the retina. It is dark

Glaucoma

The fluid in the anterior chamber of the eye, the aqueous humor, is produced by the ciliary processes. As aqueous humor is produced, it is also drained from the anterior chamber. If for some reason, the drainage system is blocked, the fluid pressure inside the eye can increase. This increase in pressure can eventually damage the retina. This increased pressure in the eye is called glaucoma.

Glaucoma is a painless condition. This is one of the reasons it is so insidious. Many people have permanent damage from glaucoma even before they know they have it. Often the visual loss in the affected eye goes unnoticed because it is compensated for by the other eye, so the patent remains unaware there is a problem until there is permanent damage.

Glaucoma can be detected by a simple measurement of intraocular pressure. This is usually done during a routine eye exam. The screening test is painless.

At present, there is no cure for glaucoma. There are treatments to help slow the progression of the disease. These include medications to lower the pressure inside the eye and interventions using a laser. In some patients there are surgical options available.

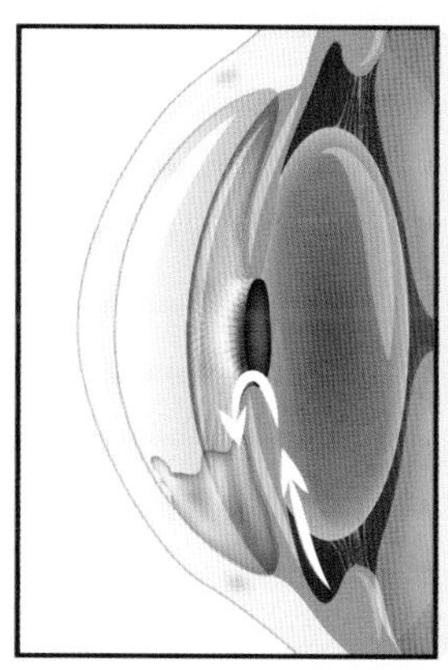

Drainage canal blocked. Too much fluid stays in the eye; this increases pressure

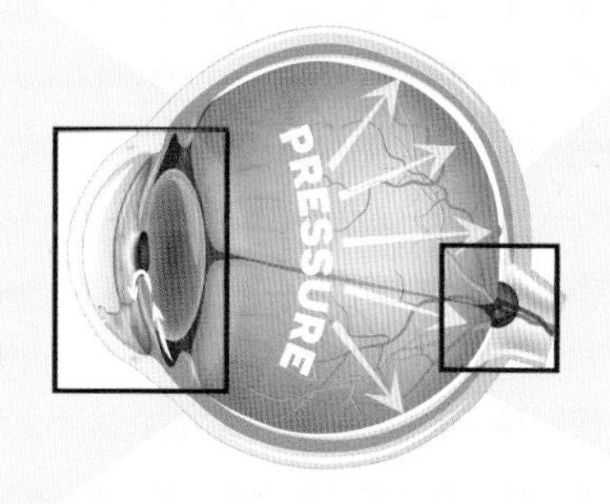

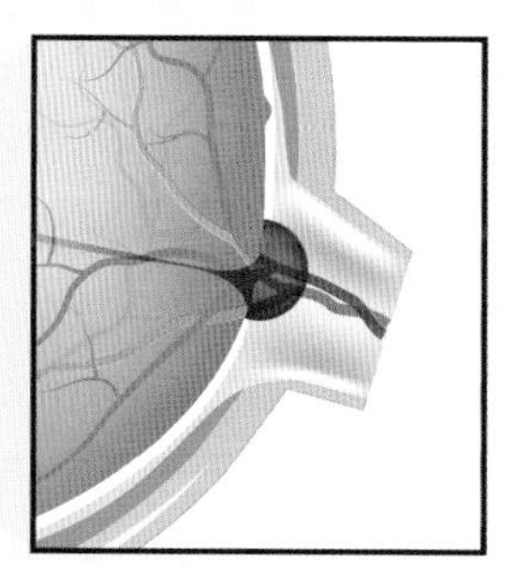

High pressure damages optic nerve.

brown in color, due to the presence of a large amount of melanin, a dark pigment. The melanin helps the choroid absorb story light rays. This helps the image on the retina to be sharp.

The front portion of the choroid becomes the ciliary body. It is made up ciliary muscles and the ciliary processes. Ciliary muscles help change the shape of the lens to improve its ability to focus. The ciliary processes secrete the fluid that fills the anterior chamber of the eye, in front of the lens.

The iris is a round, flat layer of smooth muscle with an opening in the middle. The opening is called the pupil. The iris is the colored portion of the eye. The function of the iris is to regulate the amount of light entering the eye by changing the size of the pupil.

The Retina

The inner layer of the eye is the retina. Its outer surface can be seen by looking through the pupil with an ophthalmoscope, a medical instrument that has a lens and a light source to provide a magnified view into the eye.

TAKING A CLOSER LOOK

Retina

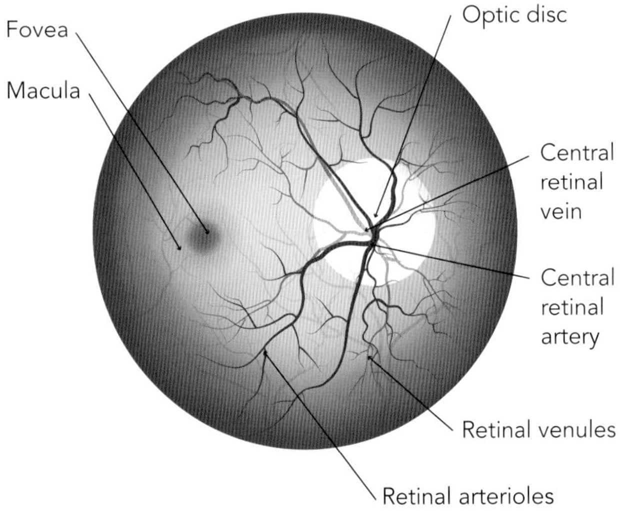

Fovea
Macula
Optic disc
Central retinal vein
Central retinal artery
Retinal venules
Retinal arterioles

The surface of the retina has several landmarks. Nearest the nose is a lighter area known as the optic disc. This is the area where the optic (II) nerve exits the eye. The optic disc is also knowns as the "blind spot." There are no photoreceptors in this area. You will also see several retinal blood vessels through an ophthalmoscope, and they tend to converge at the optic disc.

Lateral to the optic disc is a small area, essentially in the center of the retina. This is the macula lutea. In the center of the macula is the fovea centralis. The fovea is the area of highest visual resolution.

Taking a microscopic look the retina, you find an array of complicated structures. We will take them one at a time, starting from the choroid and moving inward.

Cataracts

A cataract is a clouding of the lens of the eye. As this clouding progresses, there is an increasing loss of vision. People with cataracts also have problems reading and seeing at night and often complain of having difficulty with bright lights.

Cataracts are a major cause of blindness worldwide.

Cataracts are primarily due to aging. Fortunately, the cloudy lens can be removed surgically and replaced with an artificial lens. Lens replacement usually results in substantial restoration of visual function and an associated improvement in the patient's quality of life.

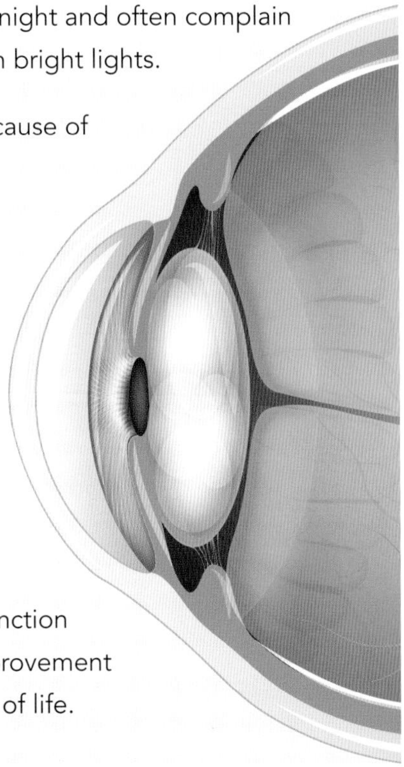

The first thing we see is the pigmented layer of the retina. This layer is made of melanin-containing cells that, like the choroid, help absorb stray light.

The next layer contains the photoreceptor cells, the rods and cones of the eye. This is where all the work is done, so to speak. Then comes a layer of bipolar cells. Then, closest to the inner surface of the retina, is a layer of ganglion cells and their axons. (The retina's two sorts of neurons are called bipolar cells and ganglion cells.)

So when you think about it, light has to go all the way through the retina—past blood vessels and past two layers of neurons—to stimulate the photoreceptor cells so that a nerve signal can make its way back out of it? Umm, well, yes.

When light enters the eye, it travels through the cornea, through the lens, through the blood vessels on the surface of the retina, through the ganglion axons, through the ganglion cell bodies, and through the bipolar cells, just to get to the photoreceptor cells. When this light stimulates the photoreceptors cells, the process of generating a neural impulse begins. That signal then begins its journey back out to the surface of the retina. It has to go to the bipolar cells, then on to the ganglion cells, then through the axons of the ganglion cells. All this to get to the optic nerve and get to the brain.

But (there's that "but" again), it's the very best way the retina could have been designed.

Is the Film in Backwards?

There are so many people in the world who mock God, so many who deny Him as Creator. So many people suppress the truth in unrighteousness.

Those who deny God as Creator must then have a way to explain all the marvelous things we see. Those people generally believe that blind evolution, millions of years of chemicals just randomly bumping together, somehow assembled all the incredibly complex things in our world. Including the human body. Including the eye.

A very common complaint against God is the design of the human eye. Many learned people have said the design of the eye proves that there is no God, or at least, if God exists He is a bad designer! After all, no engineer would design the eye in such a fashion. God put the retina in backwards, they say. The light receptors are on the wrong side. The light must go all the way to the back to be detected. A good engineer would have built the retina with the photoreceptors on the surface of the retina. So, in their view, God did it wrong.

Not only did God get it right. He got it really right! Maybe no human engineer would have designed the retina as it is, but be very glad God did.

You see, the photoreceptor cells use an enormous amount of energy in the process of converting light into nerve impulses. These cells use lots of different chemicals, need lots of nutrients, and generate lots of waste products. This high level of metabolism also generates a lot of heat that must be carried away. The design of the retina is perfect to accomplish these tasks.

The amount of blood flowing in the choroid layer is very high. It is able to provide oxygen and nutrients needed by the photoreceptors. Further, this blood flow helps dissipate the heat that is generated, keeping the photoreceptors functioning efficiently without overheating. The pigment layer also plays role in that it has mechanisms to break down harmful molecules generated by the action of light on the photoreceptors.

So, if God did put the photoreceptors in front, like so many people say He should have, there would be a huge problem. For the photoreceptors to then work properly, the pigmented layer and the choroid would have to rest on top of the photoreceptors on the INNER surface of the retina. Then how could any light get through at all?

The virtues of this great design don't end there. Scattered in these retinal layers are another kind of cell, Müller cells. Müller cells act like fiberoptic cables, efficiently transmitting the light that strikes the surface of the retina to the photoreceptor cells. In so doing, they clean up the light that has entered the eye, sharpening the images it transmits by removing any distorting light reflections and ensuring that all the colors are well-focused when they reach the photoreceptors.

I'll take God's engineering anytime!

He who planted the ear, shall He not hear?
He who formed the eye, shall He not see?

(Psalm 94:9)

Rods and Cones

Photoreceptors are the cells that convert light into nerve impulses. There are two main types of photoreceptors, rods and cones.

There are far more rods in the retina than cones. Rods are the receptors that are the most sensitive to light. Without them, we could not see in dim light or at night. Rods do not provide color vision, nor do they produce sharp images. These are found mostly in the peripheral retina.

TAKING A CLOSER LOOK

Retinal Cells

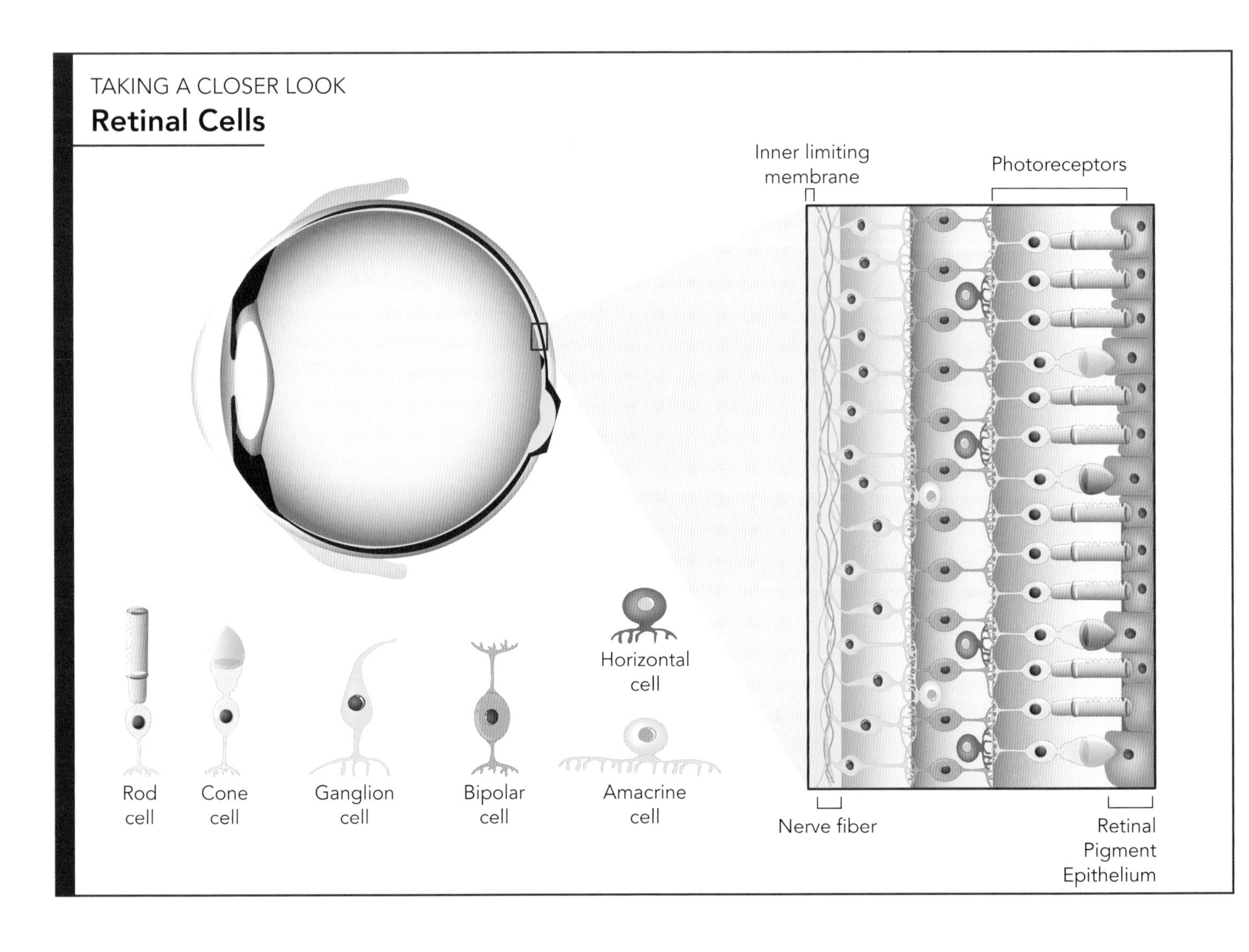

Cones are far fewer number in the retina. These photoreceptors function much better in bright light. These are more concentrated in the central retina.

Cones are also responsible for our color vision. There are three types of cones: blue cones, which are obviously sensitive to blue light, red cones, which sense red light, and green cones, which detect green light. All other colors that we perceive are mixtures of these.

How the Eye Focuses

In order for us to see things clearly, the light that enters the eye needs to be focused sharply on the retina. The focusing, or bending, of the light is done by both the cornea and the lens of the eye. The majority of the light bending, about 75 percent or so, is done by the cornea.

When light passes through a transparent object, it can be bent, or refracted. The more curved the transparent object is, the more the light is bent. Therefore, when light passes through the cornea, the curved cornea bends it. Ideally, this bending of the light places the focus precisely on the surface of the retina.

In a normal-shaped eye, with the lens in its most flattened shape (and thus having the least effect on bending the light), light from any object more than 20 feet away should focus on the retina. This point on the retina is called "the far point" of vision. Then as we progressively focus on things closer than 20 feet, the lens shape changes. It becomes more rounded and is thus able to bend the light more. This additional refraction keeps the focus point on the retina as our eye accommodates to see closer and closer objects. This is basically how the eye focuses.

But not all eyes are perfectly round.

In people who are nearsighted (like your author), the far focus point is in front of the retina, rather than on it. Their eyes are myopic. Nearsighted people can see close objects well, but have difficulty seeing more distant objects. The distant objects appear blurry or fuzzy because the light focuses too soon and then spreads out again by the time it reaches the retina.

Farsighted people have the opposite problem. In their eyes the far focus point falls behind the retina. Their eyes are hyperopic. These people can see distant objects well, but closer things appear blurry and fuzzy.

Color Blindness

Color blindness is the inability to see color or distinguish between colors. It is the result of a deficiency in one of the three types of cones. It is far more common in males. Some estimates state that as many as 10 percent of males have some degree of color blindness.

The most common form of color blindness is red-green, which signifies a lack of either red or green cones. Those with red-green color blindness see reds and greens as the same color. The second most common form of color blindness is blue-yellow.

There is no cure for color blindness. Can you tell a difference between the two squares in this illustration? If they look the same to you, you may be red-green color blind.

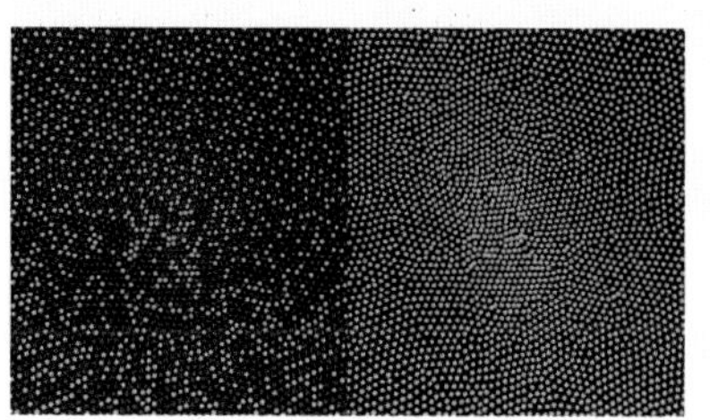

And then there is the problem that your grandparents may well experience. As people grow older, their lenses often become stiffer. Stiffer lenses cannot adjust to become rounder when looking at things that are close. Therefore, it may become difficult to read without holding a book at arm's length. This is called presbyopia.

Within certain limits, nearsightedness, farsightedness, and presbyopia can be corrected with glasses.

TAKING A CLOSER LOOK

Vision Disorders

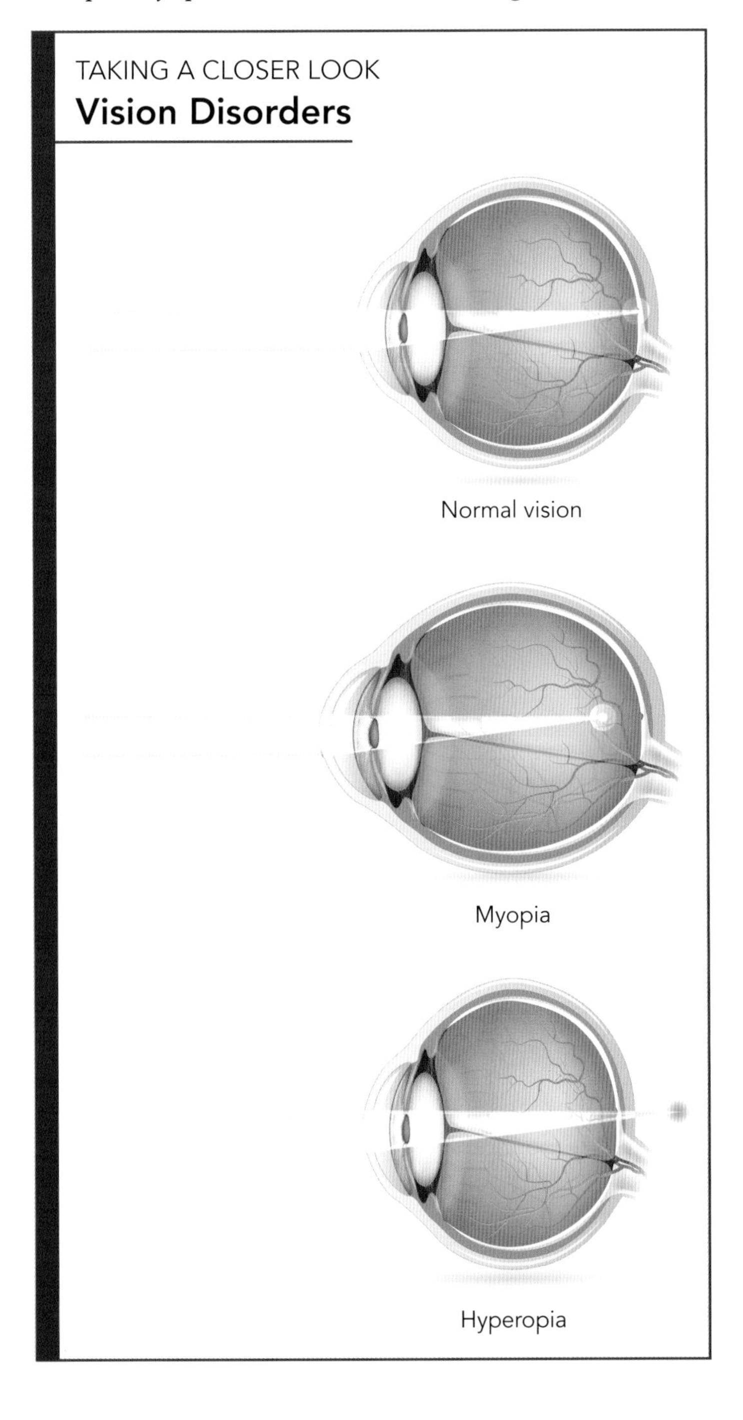

Normal vision

Myopia

Hyperopia

Epilogue

We have come to the end of our exploration of the nervous system. I am certain that because of your new knowledge about the nervous system, you now want to become a neuroscientist, or a neurologist, or a neurosurgeon! With all we know, much remains to be learned about the nervous system, so there are plenty of things left to be discovered.

At the very least, you have been introduced to the amazing complexity of the nervous system and have some understanding of the love and care used by God in its design.

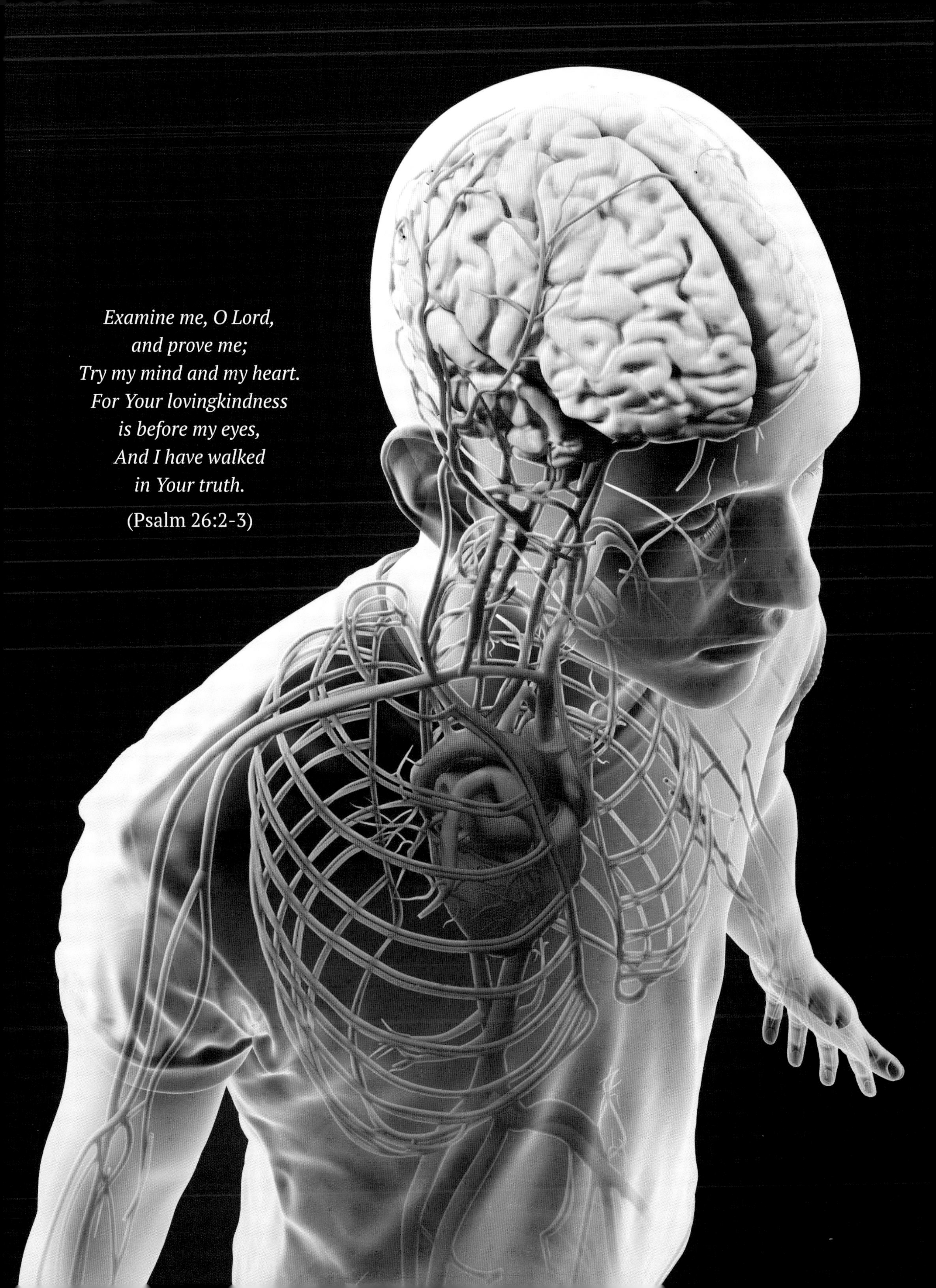
Examine me, O Lord,
and prove me;
Try my mind and my heart.
For Your lovingkindness
is before my eyes,
And I have walked
in Your truth.
(Psalm 26:2-3)

GLOSSARY

Action potential — series of events that results in a change in the membrane potential from negative to positive and back again. It is also called a nerve impulse.

Afferent — meaning "bringing toward." The sensory division of the peripheral nervous system is called the afferent division, because it brings sensory input to the central nervous system.

All-or-none phenomena — as pertains to the nervous system, there is either a full action potential or there is no action potential at all

Amnesia — loss of memory

Association areas — cortical areas of the brain that receive input from many other regions of the brain. These areas are responsible for integration of the input and determining appropriate outputs.

Autonomic nervous system — division of the peripheral nervous system consisting of motor fibers carrying impulses to smooth muscle, cardiac, muscle, and glands. Also called the involuntary nervous system.

Autonomic reflex — reflexes triggering smooth muscle and glands. With an autonomic reflex, there is no consciousness of what happened.

Axon — the portion of the neuron that carries the nerve impulse away from the cell body

Axon terminals — The most distal portion of the axon. The region where neurotransmitters are released.

Bipolar neuron — a type of neuron that has only two processes: one axon and one dendrite

Brain — one of the two parts of the central nervous system. The brain is the master control center of the human body.

Brain stem — the portion of the brain between the diencephalon and the spinal cord. It consists of the midbrain, the pons, and the medulla oblongata.

Central nervous system (CNS) — the portion of the nervous system composed of the brain and spinal cord

Cerebellum — the posterior portion of the brain. It is involved with balance and position sense.

Cerebrospinal fluid — the fluid found in the subarachnoid space surrounding the brain

Chemical synapse — a type of synapse that functions by the release and uptake of chemical messengers, called neurotransmitters. This is the most common type of synapse.

Concentration gradient — the tendency of molecules to move from areas of higher concentration to areas of lower concentration

Cones — photoreceptors in the retina responsible for color vision. There are three types of cones: blue, green, and red.

Continuous conduction — the conduction of an action potential along an unmyelinated axon. This involves step-by-step transmission along the length of the axon.

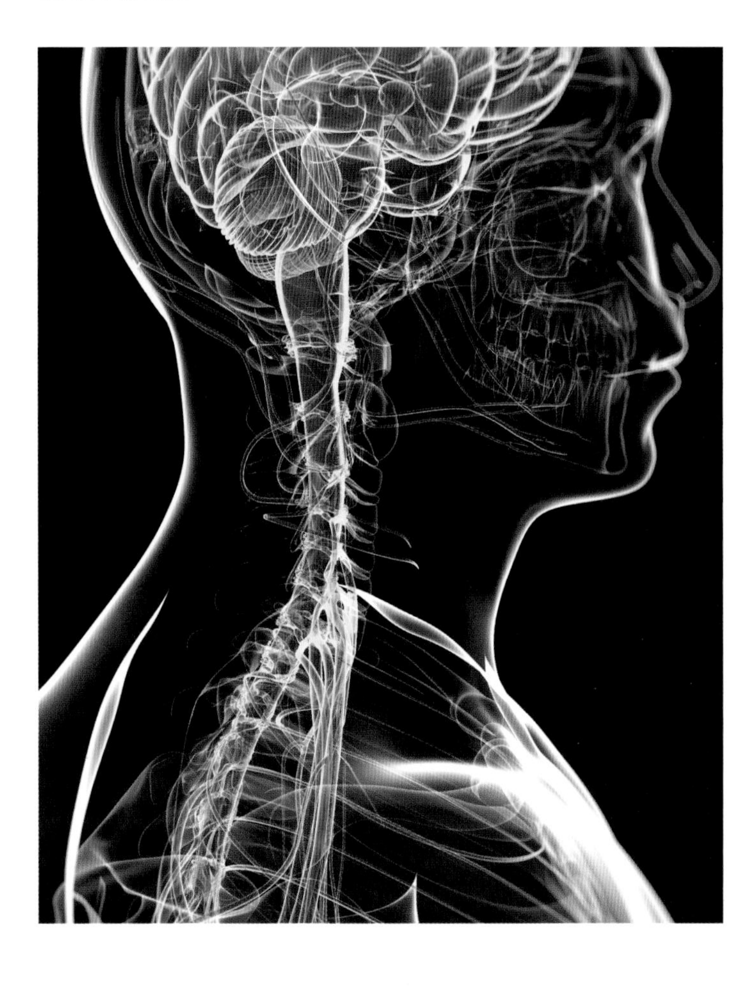

Cranial nerves — nerves that emerge directly from the brain and pass through holes in the cranium. There are 12 pairs of cranial nerves.

Decussate — crossing over to the opposite side

Dendrite —- the site where the neurons receive inputs. The signals are then carried toward the cell body.

Depolarization — a decrease in the membrane potential, where the membrane potential becomes less and less negative, and then positive

Dermatome — The region of the body that provides sensory input to a particular spinal nerve (or segment)

Diencephalon — the portion of the brain between the cerebrum and the brainstem. It consists of the thalamus and the hypothalamus.

Dorsal root — posterior root of a spinal nerve that contains only sensory axons bringing input from sensory receptors throughout the body

Efferent — meaning "carrying out." The motor division of the peripheral nervous system is called the efferent division, because it takes nerve impulses away from the central nervous system.

Electrical synapse — a type of synapse where the action potential is transmitted directly to the next cell

General senses — senses that do not require a specific sensory organ. Input for the general senses comes from basic sensory receptors located throughout the body.

Graded potential — type of membrane potentials generated in dendrites and cell bodies. This potential varies with the strength of the stimulus.

Gray matter — regions of the central nervous system consisting mostly of neuron cell bodies and nonmyelinated axons

Homeostasis — the body's ability to use many interacting mechanisms to maintain balance or "equilibrium" among its many systems

Homunculus —– means "little man." It is a mapping of the cortical regions of the brain based on the various parts of the body they either control (motor function) or receive input from (sensory function).

Hyper-polarization — an increase in membrane potential beyond the resting membrane potential

Integration — one of the functions of the nervous system. This is the recognition, analysis, and processing of various sensory inputs that results in an appropriate response.

Interneurons — neurons located in the CNS between the sensory and motor neurons. They are also called association neurons.

Lateralization — functions performed by only one of the cerebral hemispheres and not both. These functions are said to be lateralized.

Meninges — the three layers of connective tissue covering the brain and spinal cord. They are the dura mater, arachnoid mater, and the pia mater.

Mixed nerve — nerves that possess both motor and sensory fibers

Motor division — this division carries impulses from the CNS out to the body. It is sometimes called the efferent (meaning "carrying out") division because it carries nerve impulses "away from" the CNS.

Motor neuron — neurons that transmit impulses away from the central nervous system

Multipolar neurons — the most common type of neuron. It has several processes (more than three) consisting of one axon and multiple dendrites.

Myelin — a fatty substance that surrounds and electrically insulates axons of neurons

Nerves — a structure composed of many things: bundles of axons, blood vessels, connective tissue, and lymphatic vessels.

Neuroglia — one of the two types of nervous tissue. These cells support and protect neurons.

Neurons — cells that transmit electrical signals. These cells are designed to respond to some type of stimulus

Neurotransmitter — the molecules that carry the signals across the synaptic cleft

Odorant — a stimulatory chemical that triggers a smell response

Parasympathetic division — the portion of the nervous system geared to support the rest and recuperation activities of the body

Peripheral nervous system — the portion of the nervous system outside of the central nervous system. It consists of the cranial nerves that extend from the brain, and the spinal nerves that extend from the spinal cord.

Repolarization — the return of the resting membrane potential to its normal level after depolarization

Resting membrane potential — the electrical potential across the neuron membrane. It is typically around -70mV.

Rods — photoreceptors in the retina responsible for vision in dim light

Saltatory conduction — the type of nerve conduction that occurs along a myelinated axon. Here action potentials occur only at the gaps in the myelin sheath.

Sensory division — this division carries impulses from the skin and muscles as well as from the major organs in the body to the central nervous system. It is sometimes called the afferent (meaning "bringing toward") division because it carries nerve impulses "to" or "toward" the CNS.

Sensory neurons — neurons that carry impulses triggered by sensory receptors into the central nervous system

Sleep — a state in which an individual achieves a degree of unconsciousness from which he or she can be aroused

Soma — the cell body of the neuron

Somatic nervous system — the division of the nervous system that sends signals to muscles that we can consciously control

Somatic reflex — a reflex that results in contraction of skeletal muscle

Special senses — senses dependent on special types of receptors and are confined in organs specifically designed for them

Spinal cord — one of the major portions of the central nervous system. It is located in the vertebral canal.

Stimulus — a change in the environment that triggers a neuron or a receptor

Sub threshold — a depolarization not reaching the threshold level. It will not trigger a nerve impulse.

Synapse — the area where a neuron communicates with another neuron or an effector cell, such as a muscle cell.

Tastant — a stimulatory chemical that triggers a taste response

Threshold — the level of membrane depolarization at which an action potential is triggered

Unipolar neuron — type of neuron that has only one process extending from the cell body

Ventral root — anterior root of a spinal nerve that contains axons of motor neurons carrying nerve signals from the CNS out to muscles and glands

White matter — regions of the central nervous system consisting mostly of myelinated axons

Photo Credits

T=Top, M=Middle, B=Bottom, L=Left, R=Right
All photos and illustrations Shutterstock.com unless noted.

Author: Page 48, Page 110

Pubic Domain: Page 64 B

Wikimedia Commons: Images from Wikimedia Commons are used under the CC-BY-3.0 (or previous) license, CC-BY-CA 3.0, CC BY-SA 4.0, or GNU Free documentation License Version, 1.2
Page 16, Holly Fischer
Page 19, OpenStax College
Page 27, Robert Bear and David Rintoul
Page 28, Helixitta
Page 37, Page 39, Page 40, Page 41 B, Page 67, Page 73 T, Page 89 TR, Page 90 B (Blausen.com)
Page 46, CC-BY-SA-2.1-jp
Page 56 BL, Andrii Cherninskyi
Page 68 TL, Patrick J. Lynch, medical illustrator
Page 89 TL, Michael Hawke MD
Page 90 T, Welleschik
Page 101, Mark Fairchild

INDEX

DR. TOMMY MITCHELL

Dr. Tommy Mitchell graduated with a BA with highest honors from the University of Tennessee-Knoxville in 1980 with a major in cell biology. For his superior scholarship during his undergraduate study, he was elected to Phi Beta Kappa Society (the oldest and one of the most respected honor societies in America). He subsequently attended Vanderbilt University School of Medicine, where he received his medical degree in 1984.

Dr. Mitchell completed his residency at Vanderbilt University Affiliated Hospitals in 1987. He is Board Certified in Internal Medicine. In 1991, he was elected a Fellow of the American College of Physicians (F.A.C.P.). Tommy had a thriving medical practice in his hometown of Gallatin, Tennessee, for 20 years, but, in late 2006, he withdrew from medical practice to join Answers in Genesis, where he presently serves as a full time speaker, writer, and researcher.

As a scientist, physician, and father, Dr. Mitchell has a burden to provide solid answers from the Bible to equip people to stand in the face of personal tragedy and popular evolutionary misinformation. Using communication skills developed over many years of medical practice, he is able to connect with people at all educational levels and unveil the truth that can change their lives.

Dr. Mitchell has been married to his wife, Elizabeth (herself a retired obstetrician), for over 30 years; and they have three daughters. His hobbies include Martin guitars, anything to do with Bill Monroe (the famous bluegrass musician), and Apple computers. He does also admit to spending an excessive amount of time playing cribbage with Ken Ham.